Biological Invasions

Biological Invasions: Theory and Practice

Nanako Shigesada
Nara Women's University, Japan

and

Kohkichi Kawasaki
Doshisha University, Japan

This publication was supported by a generous donation from the Daido Life Foundation

Oxford   New York   Tokyo
OXFORD UNIVERSITY PRESS
1997

145769

Oxford University Press, Great Clarendon Street, Oxford OX2 6DP

Oxford New York

Athens Auckland Bangkok Bogota Bombay Buenos Aires
Calcutta Cape Town Dar es Salaam Delhi Florence Hong Kong
Istanbul Karachi Kuala Lumpur Madras Madrid Melbourne
Mexico City Nairobi Paris Singapore Taipei Tokyo Toronto

and associated companies in
Berlin Ibadan

Oxford is a trade mark of Oxford University Press

Published in the United States
by Oxford University Press Inc., New York

A catalogue record for this book is available from the British Library

Library of Congress Cataloging in Publication Data

Shigesada, Nanako, 1941–
Biological invasions : theory and practice / Nanako Shigesada and
Kohkichi Kawasaki. — 1st ed.
(Oxford series in ecology and evolution)
Includes bibliographical references and index.
1. Biological invasions. 2. Biological invasions—Mathematical
models. I. Kawasaki, Kohkichi. II. Title. III. Series.
QH353.S54 1997   574.5'247—dc20   96-34591 CIP

ISBN 0 19 854851 6 (Pbk)
0 19 854852 4 (Hbk)

Typeset by Technical Typesetting Ireland
Printed in Great Britain by Bookcraft Ltd., Midsomer Norton, Avon

Oxford Series in Ecology and Evolution
Edited by Robert M. May and Paul H. Harvey

To the memories of
Professors Ei Teramoto and
Akira Okubo

Preface

Human activities have brought about drastic changes in the global environment, one grave consequence of which is increased incidences of biological invasions. This book provides mathematical development of ideas about biological invasions on a global scale, with emphasis on deriving general biological principles from specific biological case studies.

The present book is based on an earlier one published in Japan (in Japanese) by one of the authors (N.S.), entitled *Mathematical modeling for biological invasions* (University of Tokyo Press, 1992); this English edition, however, results from the collaboration between two authors. The original Japanese edition is aimed primarily at university undergraduate and graduate students, and its purpose is to introduce to them some of the pioneering works in biological invasion as well as some mathematical models developed by the authors. Necessarily, discussions on many relevant theories, field data, and their references were curtailed in favor of conciseness. For this English edition, we have included more field data, updated some theoretical results according to the latest findings, and rewritten much of the text with a wider audience, including field researchers and scholars, in mind.

As noted in the introductory chapter, examples of biological invasions are to be found everywhere, some of the most obvious being with common plants familiar to most of us. The actual invasion process varies depending on multiple factors such as the characteristics or behaviour of the invading species, environmental conditions of the invaded site, and interactions with indigenous species. We have limited our discussion to examples that are characteristic and typical. Furthermore, most of the models we present are constructed upon simple conditions, often leading to simple, clear-cut results. Some readers may thus find that the data on invasion which they possess do not conform to any of the propagation patterns introduced in this book. It goes without saying that the living environment of organisms varies greatly, and if one is to offer a rigorous explanation of actual propagation patterns, one must prepare a model that corresponds to the detailed reality of the species in question. Even so, the authors' hopes are that many of the cases can be explained by using as the basis a model presented in this book, and modifying the model by incorporating additional factors. Indeed, if this book can provide an initial framework for those pursuing an analytical study of invasion, then our original purpose in writing it will have been met.

The authors had hoped to dedicate this book to Dr Ei Teramoto, who was largely responsible for founding and developing the field of mathematical ecology in Japan, and to Dr Akira Okubo, who wrote the classic treatise on ecological diffusion, *Diffusion and ecological problems: mathematical models* (Springer-Verlag, 1990). Sadly, both passed away at about the same time, in February of this year. The authors' relationships with Dr Teramoto go back to their student days at Kyoto University, where as their supervisor he sparked their interests in mathematical biology, leading each of them to pursue academic careers in this field. Meanwhile, each of the authors has spent a sabbatical period at Dr Okubo's laboratory at the State University of New York, where they were given opportunities to deepen their interests in the theory and practice of biological diffusion. Clearly, the authors would not be what they are today were it not for these two *sensei*s.

The authors wish to thank Dr Joel Cohen, who created for them the opportunity to publish this English version, and Dr Robert M. May, who encouraged them to publish it as part of the Oxford Series of Ecology and Evolution. Dr Cohen visited N.S. in 1990 when she was at Kyoto University, and after looking at the Japanese manuscript which she was working on at the time, encouraged her to publish a translated version. When the Japanese volume was finally published two years later, N.S. sent him a copy as a courtesy, which he obligingly forwarded to Dr May along with his letter of recommendation. This English version, then, owes its existence to these two persons. Dr May offered many valuable suggestions and comments regarding the book's composition, as well as warm support and encouragement to speed up its publication.

In the process of writing this book, the authors were fortunate to receive from many researchers in the field their latest research findings and data as well as much advice, directly and indirectly, concerning its content. These people include, from abroad, Drs David Andow, Peter Kareiva, Simon Levin, Jim Murray, Hal Caswell, Eli Holmes, Sandy Liebhold, and Hans Metz. Within Japan, many people pointed out corrections and offered constructive suggestions against the already published Japanese version. In particular, Drs Yasushi Harada, Masahiko Higashi, Kazuro Iwata, Yoh Iwasa, Masae Shiyomi, Fugo Takasu, and Nobuyuki Tuji carefully read the entire text and offered invaluable comments. Incorporating their comments, the original Japanese manuscript was extensively rewritten for the English translation, hopefully improving its readability. The authors gratefully acknowledge all those who offered assistance or support in the writing of this book, including many whose names are not mentioned here.

The authors also wish to thank Dr Toshitaka Hidaka, who originally recommended to N.S. that she publish the Japanese edition, and their editor, Mr Yoshifumi Komyo of the University of Tokyo Press. Thanks are further due to Ryu Takeguchi for his unfailing and invaluable help in producing the English translation, and to the editorial staff of Oxford University Press for their unceasing and expert support.

The authors are grateful for a generous donation from the Daido Life Foundation. They also acknowledge that their research presented in this book was supported by Grants-in-Aid for Scientific Research from the Ministry of Education, Science and Culture.

Last but not least, K.K. thanks his wife Tomoko for her wonderful personal support. N.S. expresses her gratitude to her husband, Katsuya Shigesada, for his great understanding and constant encouragements.

September 1996

N.S.
K.K.

Contents

1

Introduction

In nature, all organisms migrate or disperse to some extent. This can take a diversity of forms as in walking, swimming, flying, or being transported by wind or flowing water. Among organisms that have the ability to move, many possess the habit of moving back and forth within a definite range on a regular basis. For instance, zooplankton repeat diurnal vertical movements in which they float to the water's surface during the daytime, while sinking to the bottom at night. Many birds and mammals return to their nests or dens at the end of an active day spent in their respective home range. The most commonly identified return migrations are those with a yearly periodicity. Migratory birds such as swallows return regularly to the same breeding area year after year. Some fish and whales migrate between the equatorial zone and polar regions in their search for feeding or breeding grounds. However, a species' range cannot expand beyond fixed ranges by such return migrations alone. It can expand only if there are certain individuals who disperse to new areas without returning to their original places.

Dispersive movements become noticeably active when an offspring (or seed) leaves its natal sites, or when an organism's habitat deteriorates from overcrowding. Many marine organisms, such as reef-dwelling shrimp or shellfish that spend their larval period floating in the open seas, and most insects have within the individual's life cycle a built-in period of dispersion. Among vertebrates, exploratory migration is a characteristic of the dispersal patterns of immature animals, often called post-juvenile dispersion. If a new suitable habitat is found by such dispersal, the range can expand; if not, the dispersion ends up being wasted. 'Invasion' occurs when a species colonizes and persists in an area which it previously had not inhabited.

In an area where the environment remains virtually unchanged and which is isolated from the external world, inhabiting organisms are considered to expand their range as much as their dispersive capacities allow, after which a stable, balanced state is maintained. In reality, however, not only is an organism's living environment spatially and temporally changing, it is also an open system into which various species are moving.

Seen from a geological time scale, the geographical distribution of species on the earth's surface has changed each time a large-scale climatic or geomorphological change has taken place (Cox and Moore, 1993). These changes have resulted in geographical separations in a species' range, at times causing further speciation. For example, with glacial expansions the flicker, which inhabited the central forests in North America, shifted its range southward and spent the glacial period as two separate groups, east and west, divided by the Rocky Mountains. Eventually, when the climate became warmer again, the two groups moved back north to meet once again, but they had become different enough from each other so as to be classified as distinct species (Udvardy, 1969).

As an example of a large-scale invasion caused by geographical changes, fossil records show that when an upheaval at the present Panama canal site bridged the North and South American continents in the Pleistocene epoch some two million years ago, it caused organisms from both sides to intermix with each other. Because they had not been directly connected for a long period, the two continents shared no mammalian species that belonged to the same family groups; once the Panama land bridge initiated faunal mixing, the number of families on each continent increased drastically. Due to the competition between existing and invading species, however, many of them subsequently became extinct, and in the biota that eventually resulted after reorganization, the number of families had fallen to about the same level as before intermixing had occurred (25 in North America, 30 in South America) (May, 1978; Cox and Moore, 1993). This is often cited as supporting evidence for the dynamic-equilibrium theory of MacArthur and Wilson (1967), which states that in a certain area (particularly an island) the number of invader species and of those that become extinct balance each other, thus achieving equilibrium in the number of species. Most South American species that moved into North America eventually died out, which is thought to be the result of the superior competitive strength of North American species. This is explained by the fact that the North American continent had previously been connected to other continents in the northern hemisphere, making its ecosystems less susceptible to invasion by alien species than in South America.

More recently, when Krakatau, a volcanic island located between Java and Sumatra, erupted in 1883, the biota on the island was wiped out under a rain of hot volcanic ash. Systematic studies of the subsequent rehabitation process have obtained valuable data on invasion. These results show that the island was recolonized by plants and animals from the adjacent lands, and after fifty years had already a rich and maturing jungle of forest inhabited by epiphytic plants and many kinds of animals; particularly, the number of bird species has already reached equilibrium. However, the number of plant species is still growing and outbreaks of certain insect species often occur, showing that even 100 years later a stable, steady state has not been reached (Elton, 1958; Simkin and Fiske, 1983).

The examples given so far are of invasions accompanying environmental fluctuations caused by nature. When the Quaternary glacial period ended and mankind arrived on the scene, human activities began causing various organisms to be transported to new areas, sometimes purposefully, at other times inadvertently. With the European discovery of the New World (1492) and its subsequent colonization, many Old World plants and animals invaded the Americas or Australia and rapidly drove native species to extinction. In one episode, when Captain Cook revisited the islands of New Zealand four years after his first landing in 1769, the botanist who was accompanying his voyage found to his surprise that Canary grass, a plant native to the Mediterranean, had already established itself in several places (Crosby, 1986). As another notable example, in the tropical insular fauna of Mauritius, prior to colonization in the 17th century, there were at least 23 taxa of endemic landbirds, 12 reptiles, and two fruit bats. Currently only nine endemic landbirds, four geckos, one skink and one fruit bat survive on the mainland of Mauritius (di Castri, 1989).

A hundred years of faster and bigger transport has kept up and intensified this bombardment of every country by foreign species, brought accidentally or on purpose, by vessel and by air, and also overland from that used to be isolated.... The real thing is that we are living in a period of the world's history when the mingling of thousands of kinds of organisms from different parts of the world is setting up terrific dislocations in nature. We are seeing huge changes in the natural population balance of the world.

The above excerpt is from Elton's *The Ecology of Invasion by Animals and Plants* (1958). Even though some 40 years have passed since the publication of this classic work, this description still retains its original relevance. Although most countries today practise a quarantine system as a line of defence against biological invasions, more and more people are travelling and intermixing on an international scale and this creates increasing opportunities for invasions by organisms or diseases. There is also the danger that new life forms which were created artificially through bioengineering may escape from the laboratory and spread. Meanwhile, artificially disturbed areas, such as forests and rivers destroyed by human activities or cultivated land and pastures, are rapidly increasing, and this has increased the number of cases in which organisms which had hitherto been unsuccessful are now succeeding in invading and inhabiting new areas.

As we have seen so far, the issues of invasion and propagation are not only of academic interest but have been age-old concerns to humans because of their relevance to human society. Particularly with the accumulation of quantitative data in recent years, momentum has been building to obtain a mathematical understanding of these issues.

Much of the literature on invasion has dealt with three themes (Roughgarden, 1986; Williamson, 1989; Hengeveld, 1994):

(1) the conditions necessary for an invasion to take place,
(2) the way the invasion progresses through space,
(3) the properties of the fauna that is assembled by successive invasions.

The main goals of these works are to predict:

(1) which species will become an invader;
(2) what kind of habitat is susceptible to invasion by a particular species;
(3) if an invasion occurs, how fast it will spread;
(4) after an invasion has spread, what biological impact the invader will have on the native biota.

Recently, these questions have actively been addressed by many investigators either individually or in groups for a variety of ecosystems seen worldwide (see the publications, asterisked in References, of the SCOPE Program on the ecology of biological invasions). Among such efforts, the mathematical modelling of the spatial spread of invaders is a relatively well-developed area which has been extensively tested as well. In this book, instead of treating invasions as local events, we look at the process of invasions occurring on a global scale, and examine how mathematical models have been constructed and applied to understanding the various aspects of range expansion of invading species.

As we introduce in the next chapter, mathematical modelling of the spatial propagation of invasions was initiated by J. G. Skellam in 1951. By using the diffusion equation combined with population growth, he discovered that the range front of an invader species advanced at a constant velocity. This result was subsequently found to apply to many invader organisms, and so research in this field is widely considered as one of the most successful examples of modelling. However, some recent data have uncovered cases for which this principle does not apply. For instance, there are cases when movement takes place not only by random diffusion but also through long-distance dispersal or when environmental conditions of the invaded area are constantly changing in both time and space; such cases cannot be treated within the framework of Skellam's model. This book attempts several new approaches to such problems.

First, in Chapter 2, we introduce typical spatial expansion patterns for some well-documented cases of invasion. In Chapters 3 to 5, we discuss mathematical models that explain the characteristic features of these expansion patterns and explore the mechanisms by which invasion progresses. In particular, Chapter 3, centring on Skellam's model, will discuss the expansion pattern caused by random diffusion associated with repro-duction. Chapter 4 derives conditions for successful invasions when the environment is changing in patchwork fashion, and also attempts applica-tions to environmental issues. Chapter 5 develops a stratified diffusion model which describes invasions by organisms that extend their range by random diffusion as well as long-distance movements. Chapters 6 and 7 deal with cases when invading and native species compete for space, and show how the range of native species is pushed back by the invasion. In particular, Chapter 7 deals with disturbances that cause the environment to change in both time and space, and explores principles of coexistence between the invading and native species. Chapter 8 presents several analyses on the invasion of a predator or parasite, as well as a model of biological control by sterile insect release. Chapters 9 and 10 turn to invasions and the spread of diseases. We first introduce basic theories of epidemiology and then apply them to epidemics, such as measles and bubonic plague, among human populations. We also discuss the effects that diseases have had on human demography. Chapter 10 discusses the spread of rabies currently occurring in Europe; we will make predictions on the periodic occurrences of its epidemics and on their rate of expan-sion, and also discuss measures to control its spread.

Since this book's primary purpose lies in explaining how to construct models, the presentation of mathematical derivations from the models has been minimized, while diagrams have been used to maximum effect to obtain an intuitive understanding. For those readers interested in the analytical methods of mathematics, sections presenting simple derivations of the major equations are placed at the end of some chapters. By doing so, we hope to make this book largely self-contained.

2

Invasion of alien species

2.1 Mammals

We begin with a classic example of invasion, the spatial spread of muskrats (*Ondatra zibethica*) in central Europe from 1909 to 1927. Figure 2.1, which appeared in Elton's book (1958), gives the contours of the range of spread based on Ulbrich's data (1930). As will be shown in section 3, Skellam (1951) analysed the data using the diffusion model, laying the foundation for the theoretical study of biological invasion in an ecological context.

The muskrat, which is a species native to North America, was brought to Europe for fur-breeding. In 1905, five muskrats escaped from a farm located near Prague in Czechoslovakia. They started to spread and reproduce (later mixing with other escaped groups), inhabiting the entire European continent in the short period of 50 years; today they number many millions.

Because the wild mustrats destroy roads and dikes and cause extensive damage to agriculture, campaigns to fight their spread have been actively conducted in many European countries. Eradication campaigns led to successful results in Britain (in the 1930s) and in Scotland (Usher, 1989), whereas their control has been unsuccessful in continental Europe. In some countries, the eradication programme keeps track of all the catches in each municipality making provincial counts available, from which various demographic invasion parameters have been estimated.

When Skellam (1951) calculated the area of the muskrat's range based on the map by Ulbrich, took its square root and plotted it against years, he found that the data points lay on a straight line, as in Fig. 2.1(b) (more detailed analyses have been carried out by Williamson and Brown, 1986; Hengeveld, 1989; Andow *et al.*, 1990, 1993; see also sections 2.6 and 3.7). If the range is expanding in approximately concentric circles, with a common centre at the point where the muskrat originally escaped, the square root of the range area corresponds to the effective radii (or, to be accurate,

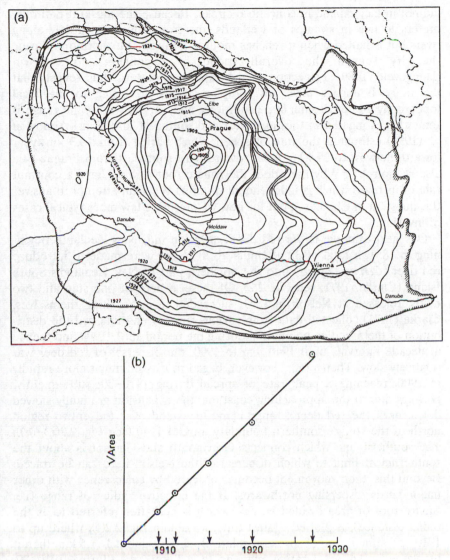

Fig. 2.1 Spread of muskrat in Europe: (a) range expansion of muskrat from 1905 to 1927 (after Elton, 1958); (b) Square root of area occupied by muskrat versus time (after Skellam, 1951).

radii times $\sqrt{\pi}$) of the circles. (There have been many other attempts to measure the range distance, which are summarized in section 2.6.) Thus, the results showed that, on average, the front of the muskrat's range advanced at a constant speed. As seen in Fig. 2.1(a), however, the range did not in fact expand in concentric circles. Because the muskrat naturally prefers to live in swamps or wetlands, the front advances rapidly along rivers but is halted when it reaches mountain ranges. Furthermore, during the dry season the overall speed of expansion slows down (Hengeveld, 1989). Inspection of the expansion pattern thus reveals that the range boundary takes on complex shapes according to the local topography and environmental conditions, and, on the whole, the range is somewhat elongated in the northwest and southeast directions. Andow *et al.* (1990) estimated the rate of speed, which varied from 10.3 km/year toward the west to 25.4 km/year towards east-southeast from Prague (see also section 3.7). Why then does the square root of area have a constant rate of increase which is independent of time? We shall attempt to answer this question in Chapter 4, but for now we present a few more typical cases of invasion.

Our next example is the red deer's invasion into New Zealand. Beginning in 1851, a total of 32 attempts at 20 places were made to introduce red deer (*Cervus elaphus*) into the northern part of New Zealand's South Island (Clarke, 1971; see Fig. 2.2). Of these, a herd of one stag and two hinds, released at Nelson in 1861, succeeded in establishing themselves. Clarke (1971) collected data from various sources and mapped the distribution of the breeding population during the period 1861–1900 and thence at decade intervals until 1940. Up to 1900, the dispersal of red deer was relatively slow. Their range, however, began to expand much more rapidly in 1900, reaching a peak rate of spread during 1910–20; subsequently, perhaps due to the approaching coastline, the expansion gradually slowed down, until the red deer's range came to extend over the entire region north of the study's southern boundary around 1940 (see Fig. 2.2(a)–(e)). The southern line which connects Greymouth and Cheviot is about the southernmost limit to which dispersal of the Nelson herd can be traced. Beyond this, their movement becomes obscured by coalescence with other major herds dispersing northwards. If the effective radius of range (i.e. square root of area divided by $\sqrt{\pi}$, which is hereafter referred to as the radial distance), is plotted against time, we obtain Fig. 2.2(f), where up to 1910, the annual rate of spread is relatively slow at about 1–1.6 km, then it increases in an accelerating manner to 4.3 km/year during 1910–20, ultimately entering a saturation phase with diminishing slope.

The red deer normally live in herds composed mainly of mature hinds; this group is loosely surrounded by stags and also scattered hinds with their calves, some of which are occasionally observed far outside the limits of range occupied by hinds (see 'wandering males' in Fig. 2.2). During the rutting season, however, the stags temporarily join the hind group to mate. The newborn calves are nurtured within the hinds' herd, until some of them leave when the herd's density becomes high and disperse. During

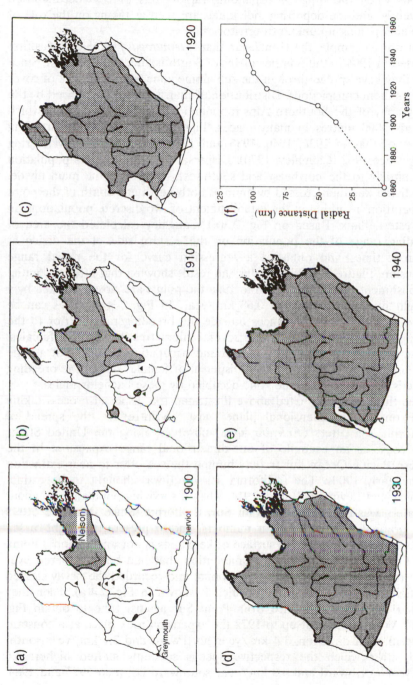

Fig. 2.2 Spread of red deer in South Island, New Zealand (after Clarke, 1971): (a)–(e) range expansion of red deer from 1900 to 1940 (▲, wandering males observed); (f) radial distance of breeding range versus time.

periods when the range is expanding rapidly this process occurs more frequently, and the departing individuals are seen as taking on the role of pioneers opening up unknown territories.

Our third example, the Himalayan thar (*Hemitragus jemlahicus*) was first released in 1904 in the Southern Alps of South Island, New Zealand. Since then they have spread through the sub-alpine zone where the lack of cover renders them conspicuous. Distribution of thar is greatly influenced by the topography of the Southern Alps mountain chain, the main divide which exceeds 3000 metres in many places. Figure 2.3(a) shows the breeding ranges of thar in 1936, 1946, 1956 and 1966 and records of wandering males since 1962 (Caughley, 1970). Dispersal from the nucleus population was initially to the northeast and southwest, parallel to the main divide. The divide was then crossed by animals both north and south of the point of liberation, resulting in the establishment of two discrete populations on its western flank. Based on Fig. 2.3(a), Caughley calculated the area of breeding range of the population on the eastern side of the divide at different times and obtained a regression curve for the radial range expansion. Figure 2.3(b) illustrates the result showing that after the initial establishment phase, from 1936 to 1966 the points are closely tracked by a straight line with a slope of 0.68 km/year. In Fig. 2.3(a), males can be found at considerable distances outside the breeding range. Prior to the mating season, some males move several kilometres in search of females. These movements result in a year-round occupation by males of a zone surrounding the breeding range. Dispersal of females out of the breeding range is necessarily into the zone occupied by these bachelor males.

The three examples cited above illustrated the dispersal process taking place on a two-dimensional plane; next we introduce the spread of California sea otters (*Enhydra lutris*) observed along the United States west coast as an example of one-dimensional range expansion. In the eastern Pacific Ocean, fur traders hunted the sea otter to near extinction in the early 1900s. The California sea otter was thought to be extinct (Lubina and Levin 1988) until 1914, when a surviving population of about 50 otters was discovered at Point Sur, California. Since then, the otters have gradually recovered their numbers, and at present, the sight of sea otters lazily napping on the surface of kelp beds is not uncommon. Lubina and Levin (1988) compiled documentation that had been gathered ever since the surviving sea otters were found, and recorded the newly sighted locations and dates on a map, which is shown in Fig. 2.4(a). Then they plotted the coastal distance from Point Sur against time to obtain Fig. 2.4(b). We can see that up to 1972 the expansion took place at a constant speed in each direction, 1.4 km/year northward and 3.1 km/year southward, after which the respective speeds suddenly shifted higher: 4.6 km/year northward and 4.8 km/year southward between 1973 and 1984.

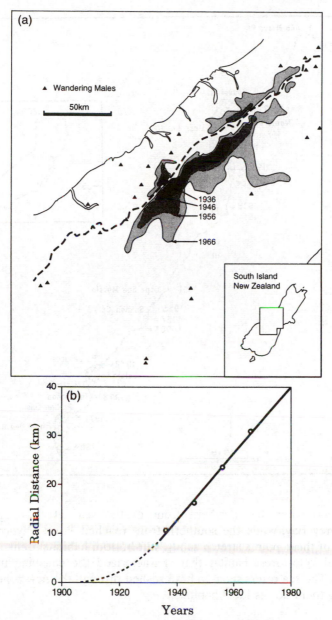

Fig. 2.3 Spread of Himalayan thar in South Island, New Zealand (redrawn from Caughley, 1970): (a) breeding range of Himalyan thar in 1936, 1946, 1956 and 1966 (▲, records of wandering males since 1962; dashed line represents the main divide); (b) radial distance of breeding range versus time for eastern population of Himalayan thar.

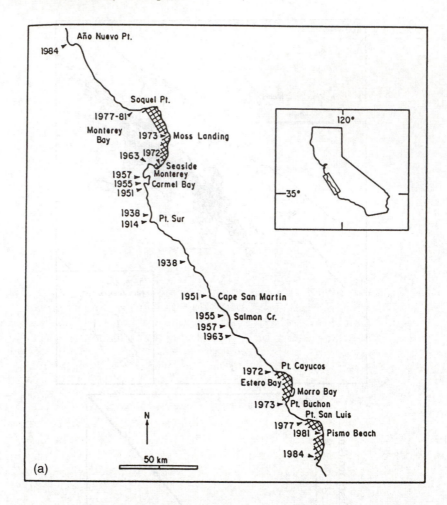

(a)

1972 was when the northern front of the sea otter's range reached Monterey Bay, while the southern front reached Point Cayucos. Beyond either of these points stretch sandy, soft-bottom habitats, unlike the rocky, subtidal kelp-forest habitat that characterized the coastline up to these points. The sea otters seem to have rushed through the new zones, perhaps finding them not as ideal habitats.

2.2 Birds

Elton (1958) states that in North America there are not many successful instances of invading European birds that have become established over

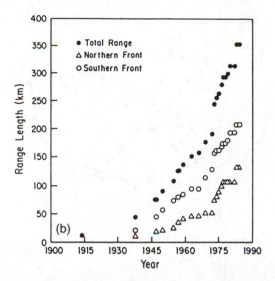

Fig. 2.4 Spread of California sea otter (after Lubina and Levin, 1988). (a) Range expansion along the central California coast. Point Sur is the traditional location of the division of range into northern and southern halves. (Cross-hatching, location of sandy or soft-bottom habitats.) (b) Range distance versus time. Range expansion was approximately piecewise linear with respect to time. There were discontinuities between 1972 and 1973 at both northern and southern boundaries, where the population fronts reached sandy or soft-bottom habitats.

the entire continent. The European starling (*Sturnus vulgaris*), the house sparrow (*Passer domesticus*), and the house finch are some of the exceptions that have been successful in doing so (Okubo, 1988).

One hundred and sixty European starlings were released in New York's Central Park in 1880 and 1881, and this was followed by several other attempts to introduce them, but it took about ten years before they established themselves. Thus, several mating pairs were found in 1895, but range expansion did not take place for a while (Wing, 1943). It was not until around 1900 that their range began expanding, eventually reaching the Pacific around 1954. Figure 2.5(a) illustrates the range observed in 1941 (Kessel, 1953; Hengeveld, 1989). The areas to the right of the bold line indicate the habitat range for breeding individuals. The black dots distributed outside the breeding range are locations where winter stragglers (non-breeding individuals without establishing wintering grounds) were sighted. They can be seen to spread out to places far removed from the breeding group and seem to act as bridgeheads from which further expansion of the breeding range occurred. Based on the data provided by Wing (1943), who traced the progress of the population front from the

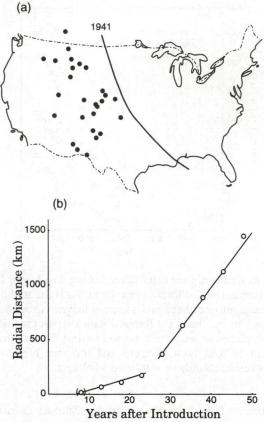

Fig. 2.5 (a) The range of the European starling in North America in 1941. The region on the right side from the bold line is the breeding range; dots indicate where young overwintering birds without territories were found (after Kessel, 1953). (b) Radial distance as function of time (after Okubo, 1988). Speed of spread was initially 11.2 km/year, then shifted to a higher constant rate, 51.2 km/year.

Christmas Bird Counts (hereafter CBCs), Okubo (1988) calculated the radial distance of the breeding range over time, which is plotted in Fig. 2.5(b). For the first ten years or so, there was an establishment phase with no discernible range expansion, and then the front advanced slowly during the next 15 years (11.2 km/year). Thereafter, the expansion rate suddenly picked up speed (51.2 km/year), which was maintained for the next 20 years or so. By the time the advancing front of non-breeding pioneer birds reached the Pacific around 1950, the range had become saturated. Wing (1943) suggested that the faster rate of spread was associated with crossing the prairie, where fewer fruits and fewer towns are available for the birds in comparison with the timbered east. An alternative explanation is presented in Chapter 5.

Compared with the European starling, the house finch was released relatively recently in the 1940s in western Long Island, New York. Beginning in the winter of 1947–48, disjunct populations in the New York City area began to appear in the reports of the annual Christmas Bird Counts. Since the CBC tradition was more firmly established than the earlier times when the starling was censused, detailed data of the winter range of the house finch were available. Based on many annual count data, Mundinger and Hope (1982) constructed pictures of range expansion for the period 1947–79. Figs. 2.6(a)–(e) illustrate the winter range expansion for several selected time intervals. The spatial change of the invaded area indicates that the house finch range expansion involved two dispersal processes: neighbourhood diffusion and jump dispersal. From the boundary of the central core area, the range reveals a gradual extension. For example, in 1962, the birds diffused northeast to link up with an outlying population in eastern Long Island, and then continued to diffuse along coastal New England during 1964–69. On the other hand, a small satellite area separated from the core range was occasionally generated by jump dispersals of a small outlying population. A typical picture of a jump dispersal is identified in the northeast direction out to the eastern end of Long Island in 1958 (Fig. 2.6(a)) and an examination of all the maps in Fig. 2.6(a)–(e) reveals many such jump dispersals. The sizes of these satellite areas increase in successive years, until they are absorbed by the main part of the range. The jump distance from the nearest border of the core range to a satellite area appears to vary. The frequency distribution of jump distance is illustrated in Fig. 2.7, from which the mean and standard deviation of jump distance are calculated as 115 km and 66 km, respectively. Overall, the house finch has radiated out in all possible directions, though there is a bias favouring movement in the southwest direction, along the Atlantic coastline, and along major river valleys. In consequence, the total area increased a thousand-fold in 30 years (1948–78) from 450 km^2 to 44×10^4 km^2.

The radial distance plotted against time for the house finch is shown in Fig. 2.6(f). To obtain a more accurate estimate of the radial distance, we calculated the radius of a semicircle which has an area equal to the range occupied by the house finch as it spread outward from Long Island (see also section 2.6). Again we find that, after the initial establishment phase (1940–47), the range began expanding at a relatively slow rate, 3.5 km/year, then switched to a faster speed, 20.7 km/year, which was maintained for the later stage of expansion during 1962–79.

Many examples of avian invasion in Europe are introduced in Hengeveld's book (1989), including the collared dove (*Streptopelia*

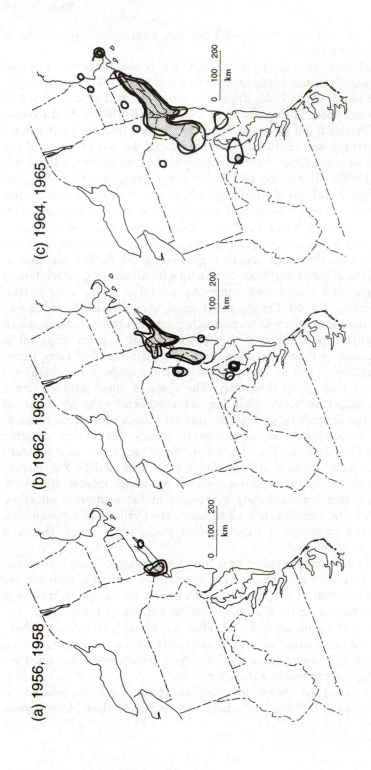

(a) 1956, 1958

(b) 1962, 1963

(c) 1964, 1965

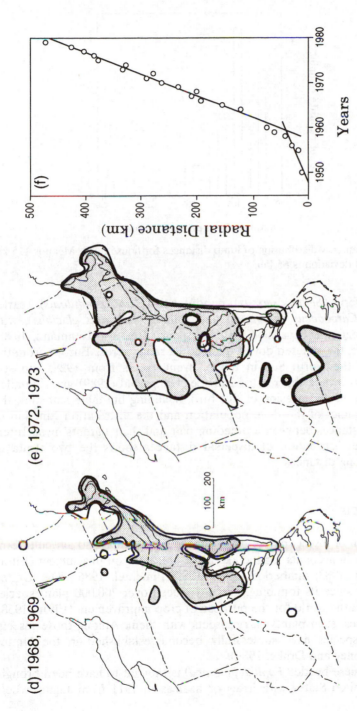

Fig. 2.6 Spread of house finch in eastern North America. (a)–(e) Expansion of winter range for several selected time intervals (after Mundinger and Hope, 1982): shaded and encircled areas represent the earlier and later years, respectively, on each map. (f) Radial distance of winter range versus time from 1950 to 1979. Speed of spread was initially 3.5 km/year, then shifted to a higher constant rate, 20.7 km/year. This calculation excludes another wave of spread that started from North Carolina around 1969 and continued to expand independently until merging with the main range originating from Long Island.

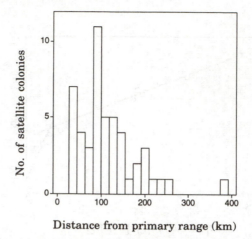

Distance from primary range (km)

Fig. 2.7 Frequency distribution of jump distances for house finch. Mean is 115 km and standard deviation is 66 km.

decaocto), serin (*Serinus serinus*), pendulin tit (*Remiz pendulinus*), scarlet rosefinch (*Carpodacus erythrinus*) and fulmar (*Fulmarus glacialis*), with maps of geographic expansion for each species (see also Isenmann, 1990). For example, the collared dove expanded its range across Europe from the Balkans to the North Sea in under twenty years from 1928, and still continues to invade the rest of Europe. Hengeveld (1989) gives detailed information on the invasion of this bird, including life-table statistics, the growth dynamics of the dove population and the distribution function of dispersal distances between a breeding pair and their parents' nest. Interestingly, the distribution of dispersal distances shows the two strata of short and long distances.

2.3 Insects

The invasion of insect species often causes great damage to agriculture or forestry, which accounts for the extensive literature on this subject (Elton, 1958; Sailer, 1983; Simberloff, 1986, 1989; Pimentel, 1986). The United States, ever since its founding, has introduced over 200,000 plant species from around the world for the purpose of crop improvement (Elton, 1958). In most cases, such plants carry insects with them. Among those insects, over 1300 species have successfully become established in the United States (Mooney and Drake, 1989).

The Japanese beetle (*Popillia japonica*) is thought to have been brought into the United States with irises or azaleas in 1911 from Japan, where they are seldom a pest. Starting from a plant nursery in New Jersey, this species spread over eastern United States (Fig. 2.8(a)) and became a major pest, not only causing damage to farm crops such as soybeans and clover to apples and peaches, but also defoliating over 250 species of trees (Elton, 1958). Looking at the radial distance plotted against time in Fig. 2.8(b), we see that spreading began at a low rate, then gradually increased its speed, and eventually reached a constant speed.

Another well-documented example of a destructive exotic organism in North America is the gypsy moth (*Lymantria dispar*), which was accidentally introduced from France to Medford, Massachusetts, in either 1868 or 1869 by an amateur entomologist. The range of the gypsy moth has expanded to most areas of northeastern North America. In addition, numerous isolated infestations arise sporadically all over the remaining areas of the United States, as seen in Fig. 2.9(a); most of these isolated colonies are either eradicated or go extinct with no intervention. Female gypsy moths in North America are unable to fly, and thus the major mechanism of dispersal is wind-borne passive movement of first instars (Mason and McManus, 1981; recently, however, a new strain has arrived from Asia, which is capable of flight). Liebhold *et al.* (1992) followed the history of the gypsy moth range expansion from quarantine data on counties designated as infested, and plotted the relationship between a county's time of first infestation and its distance from the site of the original gypsy moth introduction (Fig. 2.9(b). The data indicated that the velocity of range expansion has not been constant throughout the entire 90-year interval. Instead, there have been three distinct periods with unique rates of spread: a high rate (9.45 km/year) from 1900 to 1915, a low rate (2.82 km/year) from 1916 to 1965, and a very high rate (20.78 km/year) from 1966 to 1989 (but 7.61 km/year in counties where the mean minimum temperature was less than 7°C). Liebhold *et al.* (1992) offer possible reasons for such variable rates of expansion: thus, the slow range expansion from 1916 to 1965 was due to the 1912 enactment of the Federal domestic quarantine law against movement of gypsy moth life stages and the establishment of a barrier-zone extending from Canada to Long Island largely along the Hudson River (Perry, 1955; McManus and McIntyre, 1981). This programme was cancelled, however, in the early 1960s because of criticism against the extensive use of DDT and other substances. Another reason, besides the programme cancellation, for the rapid range expansion from 1966 to 1989 could be that forests were older and more continuous in northeastern North America during this period than the preceding 50-year period. Still another mechanism responsible for the increased rate of spread, they point out, could be accidental transportation of gypsy moth life stages by humans.

As our final example of insect invasion, we discuss the rice water weevil

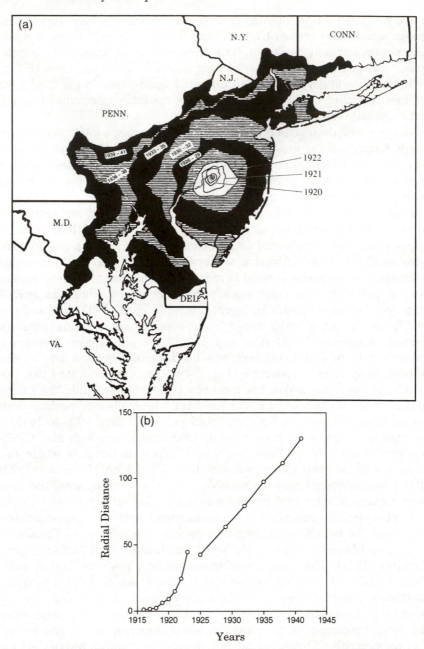

Fig. 2.8 Spread of Japanese beetle: (a) range expansion of Japanese beetle from its point of introduction in New Jersey in the United States, 1916–41; (b) radial distance of range versus time. Data from the following sources: 1916–23, Smith and Hadley (1926); 1925–41, United States Bureau of Entomology and Plant Quarantine (adapted from Elton, 1958).

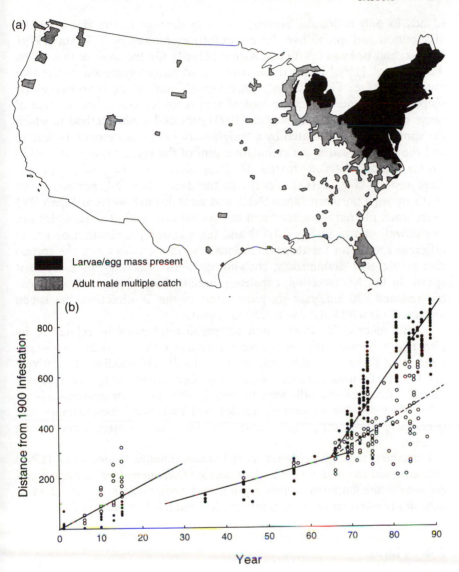

Fig. 2.9 Spread of gypsy moth in North America: (a) infested regions in 1991 (Liebhold, 1994); (b) distance from point of infestation in 1900 versus time. Each circle represents one county or census district: (●) counties with mean minimum temperature > 7°C; (○) counties with mean minimum temperature < 7°C. Linear regressions are separately drawn for 1900–15, 1916–65 and 1966–89 (after Liebhold *et al.*, 1992).

(*Lissorhoptrus oryzophilus*), which was first discovered in Japan near Nagoya city in 1976. This insect extended its region over the entire Japanese islands in only a decade, causing immense damage to the rice crop. Its distribution and spread have been carefully monitored by T. Iwata (1979), Tsuzuki and Isokawa (1976) and Kiritani (1984). On the basis of their data, Andow *et al.* (1993) mapped the progress of range expansion of the rice water weevil (see Figure 2.10a). Since Japan's main island is an irregularly shaped long island, the square root of area is inadequate as an estimate of range distance. Thus Andow *et al.* (1993) provided a new method in which the spread rate is calculated by a 'neighbourhood measurement' of spread and the total spread is the cumulative sum of the spread rate (see section 2.6 for a more detailed treatment). They divided the map of Japan into three sectors (Fig. 2.10(a))—north to the Japan Sea (N), northeast and north through northern Japan (NE), and west through western Japan (W) —and then plotted the spread and the spread rate against time, which are respectively shown in Fig. 2.10(b) and (c). Clearly, propagation occurs at different speeds for the three directions, and in each direction, the spread rate accelerated dramatically, showing no signs of settling to a constant speed. In the NE direction, expansion initially took place at 28 km/year but reached 470 km/year six years later; in the W direction, the speed increased from 47 km/year to 250 km/year.

In this species, the acceleration of spread rates could be related to its dispersal behaviour. Rice water weevil disperses by two means: by swimming from paddy to paddy, and by flying to distant paddies. If relatively few beetles fly long distances, then early observations of spread rate will mostly reflect the short-distance dispersal, while later observations will be affected by the few pioneering beetles that underwent long-distance dispersal and whose offspring became abundant enough after a certain lag of time.

Using similar neighbourhood spread measurements, Andow *et al.* (1993) also analysed the data on cereal leaf beetle (*Oulema melanopus*) and small cabbage white butterfly (*Pieris rapae*), and compared their spread rates with the predictions of some mathematical models (see section 3.7).

2.4 Plants

Reviewing the literature on invasions by terrestrial plants, Heywood (1989) stated that 'there are few ecosystems in the world that have not been affected to a greater or lesser degree by invasions by flowering plants and conifers. Human intervention has been the major causal factor in these invasions, especially through the clearance of natural vegetation for agriculture and forestry and the subsequent invasions by weedy species.'

In North America, the most successful plant invaders are annual species of Gramineae, Compositae, Leguminosae, Cruciferae, etc., which have

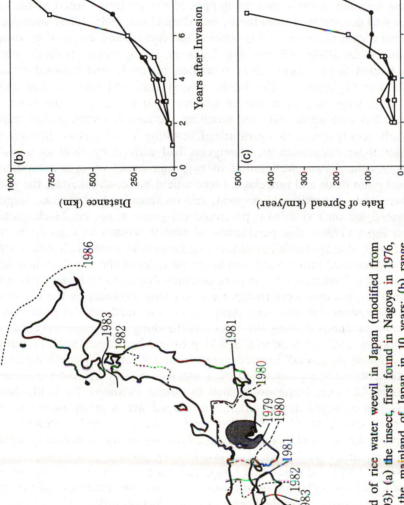

Fig. 2.10 Spread of rice water weevil in Japan (modified from Andow *et al.* 1993): (a) the insect, first found in Nagoya in 1976, spread all over the mainland of Japan in 10 years; (b) range distance versus time in three sectors; (c) rate of spread versus time in three sectors; ●, westward expansion; ○, northward expansion; □, north-eastward expansion.

mainly come from Europe, western Asia and the Mediterranean Basin. Usually the source location is climatically similar to the area of introduction, and there is likely to be similarity in the soils and vegetation (Stebbins, 1965; Baker, 1962, 1974).

Cheatgrass (*Bromus tectorum* L.) is an example of an exotic plant species whose invasion in the semi-arid steppes of the western United States has been well documented from the onset. Mack (1981, 1985, 1986) intensively collected case histories of this species and constructed maps of its range expansion for 1890–1930 (see Fig. 2.11a–e). Being remote from the eastern United States, those areas remained relatively undeveloped before 1850. But beginning in the 1850s, several gold and silver strikes were reported, triggering an inflow of settlers from the rest of the country. When they took up farming and ranching, the land was subjected to major disturbances (particularly overgrazing), resulting in soil surface alterations. Under these circumstances, cheatgrass had arrived by 1889 as a grain contaminant. It was well suited to two main types of agro-ecosystems, cereal grain fields and rangeland. Once a field was contaminated, the grass proved to be a persistent occupant, and on rangeland, cheatgrass largely replaced the once dominant perennial *caespitose* grass. As Mack quoted from Piper (1920), 'this persistence of annual bromes as a group in the seed bank of grain fields led unknowing farmers to think their wheat seed had degenerated into a weed, and in the parlance of the day the wheat had cheated the farmer'. Its seeds were not only dispersed by being carried by animal fur, but they were transported over long distances by the extensive railroad system that had been completed in the northwest by 1900 (mixed in grain or dumped along the tracks in the dung of transported cattle or used straw, etc.). As seen in Fig. 2.11 provided by Mack (1981), the range was limited to several isolated spots from 1889 to 1900. Subsequently, these spots gradually expanded their extent, while at the same time new invasion foci were being established in distant locations. By 1930, cheatgrass had occupied its current range, shaped like a group of coalescing 'leopard spots'. Mack (1981) also plotted the estimated occupied area against time for 1900–30 (see Fig. 2.11f). As in the previous example, range expansion took place at an accelerating pace.

Baker (1986) also noted that range expansion by two modes—the steady advance of a population, on the one hand, and the scattering of satellite populations from an original centre of introduction followed by a filling in of the gaps, on the other—is the most frequent pattern of plant invasions in North America. Further detailed surveys of established alien plants in North America and Hawaii were given by Mooney and Drake (1986). (Also see Kornberg and Williamson (1986), di Castri *et al.* (1990) and Groves and di Castri (1991) for plant invasions in Europe and the Mediterranean Basin, and Ramakrishnan (1991) for those in the tropics.)

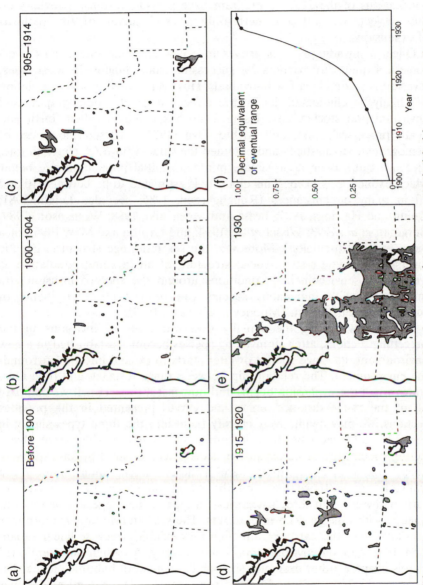

Fig. 2.11 Spread of cheatgrass in western North America (after Mack, 1981): (a)–(e) range expansion from 1890 to 1930; (f) area occupied versus time, expressed as cumulative fraction of eventual range.

2.5   General features and classification of expansion patterns

There have been a number of attempts to characterize general patterns and processes of successful establishment. Since many articles including publications of the SCOPE Program have provided extensive reviews on this subject, we will give here only a brief survey of the patterns of expansions.

Once an invading species arrives in a new environment, it must overcome a sequence of barriers for successful establishment. In most cases, invasion is started by a few individuals. However, a small initial population will likely be eliminated due to the effects of genetic, demographic and environmental stochasticity. Using a stochastic model called 'birth and death processes', Goel and Richter-Dyn (1974) analysed the effect of demographic stochasticity and obtained the value $3/\ln(b/d)$ (b: birth rate, d: death rate) as an approximate minimum initial population size below which extinction is likely. This quantity is estimated to be between 10 and 20 in many circumstances (Roughgarden, 1986; see also Leigh, 1981; Quinn and Hastings, 1987; Iwasa and Mochizuki, 1988; Williamson, 1989; Burgman *et al.*, 1993; Wissel *et al.*, 1994; and Lawton and May, 1995, for a more general discussion). Moreover, the sex ratio, age structure, genetic diversity, breeding system, social structure of an invading population as well as the environmental conditions around the spot of invasion and interactions with indigenous species can be critical determinants of successful establishment (Mooney and Drake, 1989).

As we saw among the actually observed cases of invasions in the preceding sections, after an invading species becomes established in a new environment, its range sooner or later starts to expand into the surrounding environment. The relationship between range distance and time illustrates when the expansion begins and how fast it proceeds. If we compare all of the range distance-versus-time curves presented in the previous sections, we may qualitatively classify them into the three types shown in Fig. 2.12. All three types have in common an 'initial establishment phase' during which little or no expansion takes place, followed by an 'expansion phase', and a final 'saturation phase' if there is a geographical limit to the available space. If we focus on the expansion phase, the patterns are further divided into three categories. In type 1, the range always expands linearly with time, as seen in muskrats. The sea otter and gypsy moth may also fall under this category, although they exhibit piecewise linear expansion depending on environmental heterogeneity. Although not mentioned in this chapter, other examples such as the spread of bubonic plague that occurred in 14th century Europe (see section 9.4), epidemics of rabies among wildlife (see Chapter 10) and the spread of agriculture from the Middle East to Europe (Ammerman and Cavalli-Sforza, 1984) are also

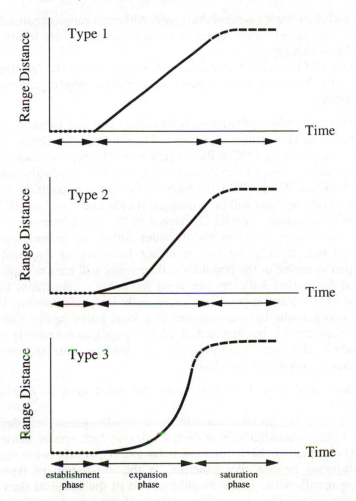

Fig. 2.12 Three types of range-versus-time curve. Range expansion commonly proceeds in three successive stages: establishment phase (dotted line), expansion phase (solid line) and saturation phase (dashed line). Expansion phase is classified into three-types. Type 1 shows linear expansion. Type 2 exhibits biphasic expansion with initial slow slope followed by steep linear slope. In type 3, rate of expansion is continually increasing with time.

known to have followed the pattern of type 1. The expansion phase of type 2 involves a slow initial spread followed by linear expansion at a higher rate. The Himalyan thar, birds such as the European starling and house finch, and the Japanese beetle are of this type. In type 3, the spread rate is continually increasing with time, showing a convex curve. The rice water weevil and cheatgrass exemplify this type. Although range expansion of the red deer falls under either type 2 or 3, the existing data are insufficient to determine which type.

There could be many reasons for the occurrence of the 'establishment phase'. The following circumstances may explain why expansion is not taking place:

1. When the invading population is extremely low and there is sufficient space for survival, travel and dispersal begin only after the home range or territory becomes filled. Particularly when the reproduction of invading species is subject to an Allee effect (reduced reproductive success at low densities: Allee, 1938), the time for the initial population density to reach a certain level will be prolonged (Lewis and Kareiva, 1993).
2. Invading organisms may be ill-adapted to the new environment so that they can barely persist at low densities within the original colonizing area. Later, if offspring with a higher fecundity or dispersal ability happen to evolve in the population, their range will start to expand. The initial lag period indicates the need for genetic adjustment before a favourable mutant appears and successfully increases (Bazzaz, 1986).
3 A few organisms initially released at a local point rapidly disperse so that they cannot be detected until the population recovers through breeding. The range expansion becomes evident when the population reaches a sufficiently high level.

A mathematical model to investigate the third case is presented in section 3.4.

The shape of the invasion curve in the 'expansion phase' may depend on the life history characteristics of each species. In fact, species belonging to each type appear to have common behavioural patterns with respect to their dispersal process. For example, in the population of type 1, the offspring usually settle in the neighbourhood of the range of their parent population (Fig. 2.13a). In contrast, species of the type 2 and 3 categories employ long-distance dispersal in addition to short-distance dispersal. In type 2, migration by short-distance dispersal expands the occupied area from its periphery, while long-distance migrants generate new satellite colonies fairly close to the primary population. Those satellite colonies will continue to expand in isolation for a short while until they eventually coalesce with their parent population (Fig. 2.13b). On the other hand, type 3 is seen when the long-distance dispersers move far away from the parent population, and over the long run expand their range independently of the other populations (Fig. 2.13c). It should be noted, however, that an alternative explanation exists for expansion curves of types 2 and 3, namely, that genetic adaptation of the invading species to its new environment leads to accelerating dispersal. The Japanese beetle and some other

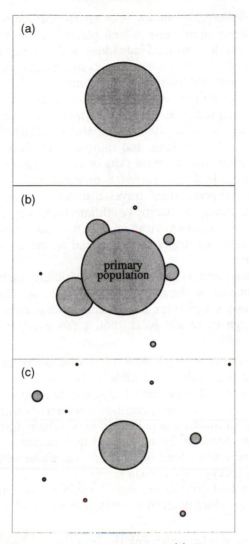

Fig. 2.13 Spatial patterns of range expansion. (a) Range expands from the periphery of the primary population. (b) Primary population expands from its periphery and at the same time satellite colonies are generated by long-distance dispersers settling fairly close to the primary population. Since both primary and satellite populations expand, their ranges coalesce before long. (c) Satellite colonies are scattered far enough from the parent population that their ranges remain isolated for a long period.

species could have involved such genetic adjustment in the early stage of the expansion phase (Hengeveld, 1989).

Species that inherently employ both short- and long-distance dispersal are not uncommon among insects and plants. In insects, there are many examples in which short-winged individuals with low mobility are produced when the population density is low, but as the population density increases or the environment otherwise worsens, long-winged varieties with long-distance flying capabilities emerge. The periodic outbreak of locusts in north Africa is a typical case. Meanwhile there are plant species known to employ strategies, such as altering the size or attached position of the seed, that will enable both long- and short-distance dispersal. Furthermore there are organisms that move on rafts of vegetation carried by rivers and marine currents, ride on air currents, or travel by attaching to birds or larger flying arthropods; their travel distance will vary by accident or according to the changing climatic conditions (di Castri, 1989). In Chapters 3 to 5, we will use mathematical models to examine quantitatively how these diverse modes of dispersal are related to the three types of range-versus-time curves.

So far we have mentioned little about invasibility related to the properties of environments or the organism's traits. Before closing this section, we present Brown's five rules concerning general patterns of successful invasion, although there are more than a few exceptions to these rules (J. H. Brown, 1989).

Rule 1: Isolated environments with a low diversity of native species tend to be differentially susceptible to invasion.
Rule 2: Species that are successful invaders tend to be native to continents and to extensive, non-isolated habitats within continents.
Rule 3: Successful invasion is enhanced by similarity in the physical environment between the source and target areas.
Rule 4: Invading exotics tend to be more successful when native species do not occupy similar niches.
Rule 5: Species that inhabit disturbed environments and those with a history of close association with humans tend to be successful in invading man-modified habitats.

Somewhat different rules which link the probability of establishment to the ecological traits of an invader, at either the species level or community level, have been presented by Mack (1981), Bazzaz (1986), Diamond and Case (1986), Lawton and Brown (1986), Crawley (1986), Pimm (1989), Mooney and Drake (1989), Ehrlich (1989), Case (1991), Lodge (1993), etc.; but see also Crawley (1987) for his pessimistic perspective.

2.6 Appendix: measurement of range distance

Range distance has been quantified in various ways depending on the measurement method. The examples presented in the preceding sections

involved three methods for evaluating this quantity: the square root of the area occupied, the linear distance measurement and the neighbourhood measurement.

The square root of the occupied area divided by $\sqrt{\pi}$ indicates the average radial distance, when the range is expanding in approximately concentric circles. Since Skellam applied this measure to the spread of muskrat, many maps of the spread of invasion have been analysed by transforming the area into the effective range radius (e.g. red deer, Himalyan thar, European starling and Japanese beetle in the previous sections). If the range is radially symmetric, this measure should provide a good estimate, even if the front shows an irregular shape. However, as noted by Andow *et al.* (1993), if the spread is impeded in some directions by geographic barriers, or little information on an organism's distribution is available for some local district (see Fig. 2.14), some modification is needed in evaluating the radial distance. For example, for case (b) or (d) in Fig. 2.14, if we measure the shaded area A and the angle θ instead of the whole area, then the shaded area is given by $A = r^2\theta/2$, where r is the radial distance. Solving this equation gives $r = \sqrt{2A/\theta}$. The radial distance for the spread of the house finch given in Fig. 2.6(f) was calculated using $\theta = \pi$.

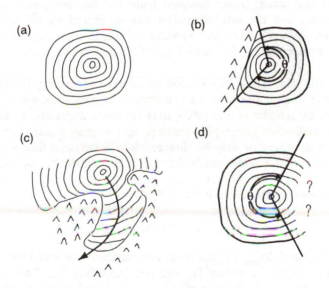

Fig. 2.14 Practical complications in the measurement of range expansion: (a) idealized spread; (b) asymmetric spread due to geographic barriers; (c) spread directed down a corridor; (d) poor information on the distribution of the spreading organism. Radial distances for (b) and (d) are calculated from $(2A/\theta)^{1/2}$, where A is the shaded area and θ is the angle of the sector (redrawn from Andow *et al.*, 1993).

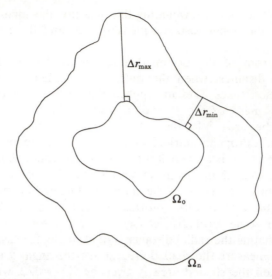

Fig. 2.15 Neighbourhood measurement of spread (after Andow *et al.*, 1993).

Linear distance measurement was used for the gypsy moth (Fig. 2.9(b)), for which the actual linear distance from the starting point was plotted against time, and the rate of spread was estimated by the slope of the regression. As an alternative measure of linear distance, we could use the distance between the initial point and farthest point of invasion (Caughley, 1970).

When the geographic distribution of an expanding population is highly asymmetric, 'neighbourhood measurements' of range expansion, recently presented by Andow *et al.* (1993), may be more appropriate. Because of habitat variation or geographic barriers, spread may occur faster in some regions than others or may be directed into a corridor (see Fig. 2.14(c)). Thus Andow *et al.* (1993) specify the rate of spread by the average of range increment in local neighbourhoods over the geographic boundary (see Fig. 2.15) and then approximate it by

$$\Delta r = \left[(\Delta r_{max}^2 + \Delta r_{min}^2)/2\right]^{1/2},$$

where Δr_{min} and Δr_{max} are the minimum and maximum distances between the old geographic boundary Ω_0 and new boundary Ω_n. The total spread is given by the cumulative sum of the rate of spread on this neighbourhood estimate. When environmental heterogeneity is separable into several sectors, this measurement has a practical advantage. As we saw in the spread of rice water weevil (Fig. 2.8), Andow *et al.* (1993) divided Japan's main island into three sectors, north, northeast and west, for each of which they estimated the spread rate and total range distance. This method was also used to re-analyse the spread of muskrat by Andow *et al.* (1990, 1993), who divided updated maps (Hoffman, 1958; Frank and Härle, 1964) into

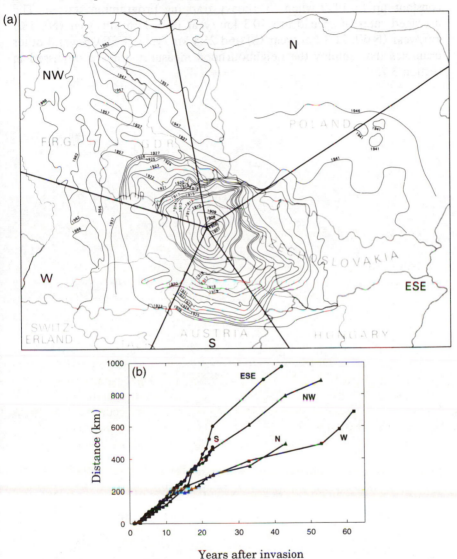

Fig. 2.16 Spread of muskrat from Prague in five sectors (after Andow *et al.*, 1990, 1993). (a) S–ESE and ESE–N boundaries roughly separate the eastern European plain; N–NW boundary distinguishes spread down Oder River from spread down Elbe River; NW–W and W–S boundaries isolate the poorly known W sector. (b) Range distance as a function of time in the five sectors.

the five sectors shown in Fig. 2.16(a) and measured the distance of spread in each sector separately. The result of the analysis is summarized in Fig. 2.16(b). The slopes of the plots indicating the spread rate are generally constant up to 1930 when the heavy trapping programme started. The averaged rates of spread are 10.3 km/year (W), 11.5 km/year (N), 18.7 km/year (NW), 21.3 km/year (S) and 25.4 km/year (ESE). Several other examples that employ the neighbourhood measurement are discussed in section 3.7.

3

Diffusion models and biological waves

3.1 Fisher's model

The spatial spread of an invading species can basically be seen as a process in which individuals disperse while multiplying their numbers. One model in which dispersal is formulated as a random diffusion process is the Fisher equation.

Assume that a few individuals invade the centre of a two-dimensional homogeneous space. If $n(x, t)$ denotes the population density at time t and spatial coordinate $x = (x, y)$, the Fisher equation in two-dimensional space is expressed as follows:

$$\frac{\partial n}{\partial t} = D \left(\frac{\partial^2 n}{\partial x^2} + \frac{\partial^2 n}{\partial y^2} \right) + (\varepsilon - \mu n)n. \tag{3.1}$$

The left-hand side indicates the change in the population density with time, which is caused by random diffusion and local population growth, expressed respectively by the first and second terms on the right-hand side. D is the diffusion coefficient, which is a measure of how quickly the organisms disperse. The population growth is formulated by the logistic growth function, where ε is the intrinsic rate of increase and $\mu (\geq 0)$ represents the effect of intraspecific competition on the reproduction rate. Fisher (1937) first proposed this equation as a model in population genetics to describe the process of spatial spread when mutant individuals with higher adaptability appear in a population. Meanwhile, Skellam (1951) applied this equation for the special case of $\mu = 0$ to the range expansion of muskrats, and ever since then it has been widely used as a fundamental equation describing the process by which a range expands, in various fields ranging from bacterial growth, population genetics and ecology to epidemiology and even the spread of human cultures.

In the following two sections, we discuss in some detail how eqn (3.1) is developed as an ecological model.

3.2 Diffusion equation

We first discuss the term in eqn (3.1) expressing diffusion. If a range expands solely by diffusion without population growth, the Fisher equation (3.1) becomes the so-called diffusion equation in two-dimensional space:

$$\frac{\partial n}{\partial t} = D\left(\frac{\partial^2 n}{\partial x^2} + \frac{\partial^2 n}{\partial y^2}\right). \tag{3.2}$$

In the field of physics, this equation has been studied extensively from quite early on to deal with a variety of processes such as heat conduction (Fourier, 1822), Brownian movement of microscopic particles (e.g. pollen particles) in a liquid (Einstein, 1905; Carslaw and Jaeger, 1959; Berg, 1983), or turbulent flow of a fluid with mixing (Okubo, 1980). It has also been used to express 'random walks' (Chandrasekhar, 1943; Renshaw, 1991) and similar movements of organisms. There are, consequently, many ways of deriving eqn (3.2), and while tracing those historical aspects can be of considerable interest in itself (see Okubo, 1980), in section 3.8 the reader will find one of the simplest derivations using the random walk model.

To solve eqn (3.2), the initial distribution must be specified. If n_0 individuals invade the origin at $t = 0$, the initial distribution is given by

$$n(x,0) = n_0 \delta(x), \tag{3.3}$$

where $\delta(x)$ is called the delta function, which indicates that the probability of finding an individual is concentrated in the immediate vicinity of the origin. Under this initial condition, the diffusion equation (3.2) is solved as:

$$n(x,t) = \frac{n_0}{4\pi Dt} \exp\left(-\frac{x^2 + y^2}{4Dt}\right), \tag{3.4}$$

which is the so-called two-dimensional Gaussian (or normal) distribution. It is easy for the reader to confirm that eqn (3.4) indeed satisfies eqns (3.2) and (3.3). If we denote by $r = \sqrt{x^2 + y^2}$ the radial distance from the origin to point (x, y), eqn (3.4) is rewritten as a function of r and t:

$$n(r,t) = \frac{n_0}{4\pi Dt} \exp\left(-\frac{r^2}{4Dt}\right), \tag{3.5}$$

where $n(r,t)$ expresses the population density of an arbitrary point on a circle of radius r. Figure 3.1 illustrates the temporal change in the radial distribution given by eqn (3.5). It shows that the distribution is expanding

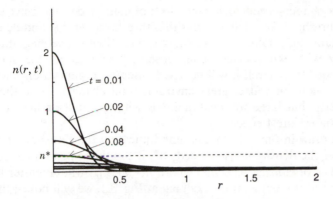

Fig. 3.1 Solution of diffusion equation (3.5). Population initially released at the origin expands radially with time. Horizontal axis indicates radius r, and vertical axis the population density at radius r and time t; $\varepsilon = 1$, $t = 0$–1. Dashed line indicates threshold density, n^*. Area where density is above n^* is defined as the range of population. Population density will eventually fall below n^* so that the range disappears.

in a concentric manner around the origin. The mean radius and mean square radius are respectively calculated as

$$\langle r \rangle = \frac{1}{n_0} \int_0^\infty m(r,t) 2\pi r \, dr = \sqrt{\pi Dt}, \qquad (3.6)$$

$$\langle r^2 \rangle = \frac{1}{n_0} \int_0^\infty r^2 n(r,t) 2\pi r \, dr = 4Dt. \qquad (3.7)$$

From the property of the two-dimensional normal distribution, 63% of the individuals are to be found within the circle with radius $\sqrt{\langle r^2 \rangle} = \sqrt{4Dt}$. In other words, the circle containing 63% of the individuals increases at a decreasing rate $\sqrt{D/t}$. If the radius is doubled to $2\sqrt{4Dt}$, then the circle will contain 98% of the population. Thus the population density falls rapidly as one moves away from the origin, but a finite value is retained out to infinity. The implication of this is that, although the probability is extremely low, some individuals will have managed to travel within a finite period to extremely distant locations; some precaution is needed on this point whenever applying the diffusion equation to range expansions in an infinite space.

Meanwhile, there are limits to our observational capacities, which should be taken into consideration. When the population density is extremely low, it becomes impossible to detect the presence of organisms because of our limited searching ability. This means that in the region where a certain organism is recognized to inhabit, its population density must exceed a certain threshold. We shall call this the 'threshold density' and denote it by n^* (Skellam, 1951; Okubo, 1980; Andow *et al.*, 1990). The magnitude of n^* will vary with the species being observed as well as the method of observation. In general, it will be small when an organism displays prominent behaviour or causes great environmental changes, as in the case of insect pests, but large for small animals who spend their lives remaining hidden under forest cover.

Let us draw in Fig. 3.1 a dotted line indicating the threshold density n^*. Whenever the density $n(r, t)$ exceeds n^*, it should be possible to detect the presence of organisms. The area where $n(r, t) > n^*$ will hereafter be called the 'range of the population'. Looking at Fig. 3.1, we will notice that if the initial population exceeds n^*, then a distribution range will initially appear about the origin; but the density will eventually fall below n^*, causing the distribution range to disappear. Thus, if an invasion occurs by an organism that spreads by diffusion alone and without population growth, its distribution range will be observed only at the initial stage.

The subject of population growth will be dealt with in the following section; here we discuss how to estimate the diffusion coefficient D, the sole parameter in (3.2). Rewriting eqns (3.6) and (3.7) gives two formulae to calculate the diffusion coefficient:

$$D = \langle r \rangle^2 / \pi t. \tag{3.8}$$

$$D = \langle r^2 \rangle / 4t. \tag{3.9}$$

If some individuals of the species are released at a certain location and it is possible to measure the spatial distribution against time, then the mean square radius $\langle r^2 \rangle$ defined by eqn (3.7) can be calculated for different times. When $\langle r^2 \rangle$ is plotted against time and the regression line determined, D can be estimated from eqn (3.9) to be one-quarter of the slope of the regression line. Similarly, using formula (3.8) provides another way to estimate D. Since measurement of the distribution pattern for a given time requires parallel observations at multiple locations, these methods normally involve a great deal of time and effort.

Alternatively, D can be estimated by tracing the path of an individual carrying out a random walk. Diffusion equation (3.2) is still available for a random walk of a single organism, although in the present case, $n(x, t)$ represents not the population density but the probability density that the organism lies at (x, y) and t. Accordingly, eqns (3.8) and (3.9) still hold if $\langle r \rangle$ is considered not as the mean radius of the radial distribution but instead as the 'mean displacement' by an individual's random walk during time t (i.e. the mean value of the straight-line distance between the starting point and point of arrival at time t). Similarly, $\langle r^2 \rangle$ represents the 'mean square displacement' by an individual's random walk during time t. To measure $\langle r \rangle$ or $\langle r^2 \rangle$, the path of small animals can be determined by direct observation or from mark-recapture data, while that of large animals can be determined by radio tracking, etc. Applying those quantities thus obtained into eqn (3.8) or (3.9) gives the diffusion coefficients, which should theoretically be identical for both methods. It is also important to check whether D, derived in this manner, remains constant regardless of time t. If so, it shows that the movement of the organism being observed is consistent with the diffusion model.

There have been many attempts to measure the diffusion coefficient from field data of animal tracing. If mark-recapture data are available, eqn (3.8) is approximated as

$$D = \frac{\left(\dfrac{\displaystyle\sum_{n=1}^{N} (\text{distance of recaptured organisms from release site})}{\text{total number of recaptured organisms}} \right)^2}{\pi \times (\text{time from release})}. \tag{3.10}$$

With the use of eqn (3.10), Kareiva (1983) determined D for 12 herbivorous insect species, and then substituted their values into eqn (3.4) to construct spatial distribution maps. He found that, in eight of the species, the map agreed well with the actual distribution. He notes, however, that diffusion coefficients vary significantly both within and between species depending on local habitats, which may be caused by species-specific responses to habitat or resource patchiness.

Equation (3.10) was also applied to the collared dove (*Streptopelia decaocto*) by Holmes (1993) and to the cabbage white butterfly (*Pieris rapae*) by Andow *et al.* (1990, 1993). The values of D thus estimated for various species are summarized in section 3.7.

3.3   Logistic equation

In the previous section, we considered the spread of population by diffusion alone. If conversely the population changes by growth (i.e. reproduction) alone without diffusion, eqn (3.1) becomes the so-called logistic equation:

$$\frac{dn}{dt} = (\varepsilon - \mu n)n, \tag{3.11}$$

where $n(t)$ is the population density at time t for some fixed location. $(\varepsilon - \mu n)$ represents the per capita growth rate, which declines linearly with the density. The intrinsic rate of increase ε is the growth rate (i.e. difference between birth rate and death rate) when density is low, while μn represents the density effect on the reproductive rate. As the density grows, competition for food or space increases, either directly through interactions between individuals, or indirectly by exploitation of resources, resulting in the decline of the rate of reproduction. Conventionally, eqn (3.11) is written, by putting $K = \varepsilon/\mu$, as

$$\frac{dn}{dt} = \varepsilon\left(1 - \frac{n}{K}\right)n, \tag{3.12}$$

where K is called the 'carrying capacity'.

If there is no competition within the species (i.e. $\mu = 0$), the logistic equation becomes the so-called Malthusian equation:

$$\frac{dn}{dt} = \varepsilon n, \tag{3.13}$$

which is easily solved as

$$n(t) = n(0)e^{\varepsilon t}, \tag{3.14}$$

where $n(0)$ denotes the initial density. When ε is positive, the population increases exponentially without limit.

On the other hand, when competition exists within the species (e.g. $\mu > 0$), the solution for eqn (3.12) is given as follows:

$$n(t) = n(0)\frac{Ke^{\varepsilon t}}{K + n(0)(e^{\varepsilon t} - 1)}. \tag{3.15}$$

Figure 3.2 shows the change in density over time as given by eqns (3.14) and (3.15). Initially, when the density is low, the curves for both equations increase exponentially; with increasing density, the effect of competition

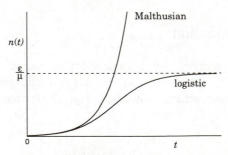

Fig. 3.2 Solutions of Malthusian equation and logistic equation.

becomes apparent in the logistic equation, with the growth rate slowing down after the density reaches half the carrying capacity, and eventually the density asymptotically approaches the carrying capacity, K.

Because of its simple structure and explicit solution, the logistic equation has been widely employed in theoretical work and in empirical studies to describe the growth of populations both in the field and under laboratory conditions (D. Brown and Rothery, 1993). For example, Hengeveld (1989) showed that the number of muskrats caught in successive years in the Netherlands shows an S-shaped curve conforming to logistic population increase.

3.4 Skellam's model

In the previous two sections, we examined the respective effects that diffusion and reproduction have on distribution and population dynamics. We are now ready to use eqn (3.1), with both factors involved, and see how the distribution spreads.

Since Fisher (1937) proposed eqn (3.1) (in reality, he investigated a one-dimensional equation that considered diffusion only along the x-axis), it has been extensively studied by Kolmogorov *et al.* (1937) and many other mathematicians (e.g. Kametaka, 1976; Bramson, 1973; Fife, 1979; Britton, 1986; Murray, 1989). Since a full discussion of their work would require an extensive background in mathematics, here we introduce the intuitive approach originally employed by Skellam (1951) and Fisher (1937), incorporating some recent results presented by Andow *et al.* (1993) and Shigesada *et al.* (1995).

As a special case of eqn (3.1), Skellam analysed the spread of a population when growth is Malthusian. We first introduce his treatment for its simplicity and its clear-cut results. The diffusion equation with Malthusian growth is obtained by letting $\mu = 0$ in eqn (3.1):

$$\frac{\partial n}{\partial t} = D\left(\frac{\partial^2 n}{\partial x^2} + \frac{\partial^2 n}{\partial y^2}\right) + \varepsilon n. \qquad (3.16)$$

This equation is hereafter referred to as the Skellam model. Multiplying both sides of eqn (3.16) by $\exp(-\varepsilon t)$ and rearranging the terms, we have

$$\frac{\partial n e^{-\varepsilon t}}{\partial t} = D\left(\frac{\partial^2 n e^{-\varepsilon t}}{\partial x^2} + \frac{\partial^2 n e^{-\varepsilon t}}{\partial y^2}\right). \qquad (3.17)$$

Since this equation has the same form as the diffusion equation given by

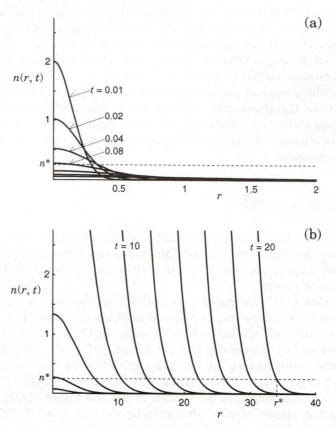

Fig. 3.3 Solution of Skellam model, with parameters $\varepsilon = 1$, $D = 1$. (a) Change in distribution for $t = 0$–1. Spatio-temporal pattern is almost the same as that from the diffusion equation given in Fig. 3.1. (b) Change in distribution for $t = 2$–20. The front of the distribution tends to advance at constant velocity while maintaining its shape. Dashed line indicates threshold density, n^*; r^* represents range radius of population at $t = 20$. (Shigesada *et al.*, 1995)

(3.2), the solution, which is subject to the initial condition (3.3), is given by eqn (3.5) multiplied by $\exp(\varepsilon t)$:

$$n(r,t) = \frac{n_0}{4\pi Dt} \exp\left(\varepsilon t - \frac{r^2}{4Dt} \right). \qquad (3.18)$$

Figure 3.3 shows how this distribution varies with time. The distribution at any given time is Gaussian, as is the case for the diffusion equation. In particular, the spatial pattern of Fig. 3.3(a) is quite similar to that of Fig. 3.1, so that the growth term does not seem to affect the distribution pattern for some initial period. If we look at some fixed location, the density increases with time in an exponential manner when t is large. Thus the population range continually expands due to the combined effects of diffusion and growth, and its front seems to advance at a constant velocity while maintaining its shape.

How fast does this wave front move? If, as in section 3.2, n^* denotes the threshold density below which the population cannot be detected, the distribution range of that organism is those areas where density $n(r,t)$ exceeds n^*. Letting r^* denote the position at which $n(r,t)$ intersects the threshold level n^*, then r^* corresponds to the front of the distribution range, or the radial distance. So by letting $n = n^*$ and $r = r^*$ in eqn (3.18), and solving r^* as a function of time, we obtain the following equation which shows the relation between the range distance and time:

$$r^* = 2\sqrt{\varepsilon D}\ t\left(1 + \frac{1}{\varepsilon t}\log\frac{n_0}{4\pi Dtn^*}\right)^{1/2}. \tag{3.19}$$

While our goal is to establish this equation's properties, the many parameters it contains can be somewhat distracting. To make eqn (3.19) simpler we first carry out on r^* and t the following scale conversions:

$$R^* = \sqrt{\frac{\varepsilon}{D}}\ r^*,\ T = \varepsilon t. \tag{3.20}$$

R^* and T are quantities that are proportional to the range distance r^* and time t, respectively, and since they are both non-dimensional, they may be called the non-dimensionalized range distance and time, respectively. Substituting eqn (3.20) into (3.19), we obtain the following relation between R^* and T:

$$R^* = 2T\left(1 + \frac{1}{T}\log\frac{\gamma}{4\pi T}\right)^{1/2}, \tag{3.21}$$

which contains a single parameter,

$$\gamma = \frac{\varepsilon n_0}{Dn^*}. \tag{3.22}$$

The curves showing the relation between R^* and T are illustrated in Fig. 3.4 for different γ values. In (a), where $\gamma = 10$, the range distance R^* increases monotonically from the outset and asymptotically approaches a straight line with a slope of 2. In (b), where $\gamma = 2.5$, a small distribution range initially appears around the invasion point, only to disappear immediately, followed by a period during which R^* remains zero. This period lasts a while, after which R^* begins to increase from zero, eventually

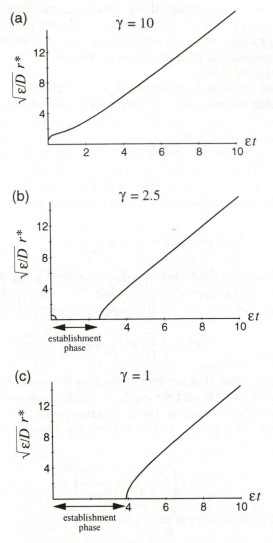

Fig. 3.4 Range distance versus time for different values of γ. Vertical axis indicates R^* ($= \sqrt{\varepsilon/D}\, r^*$) and horizontal axis, T ($= \varepsilon t$). Slope asymptotically approaches 2 for any value of γ ($= \varepsilon n_0/Dn^*$). Establishment phase appears when γ is less than $4\pi/\mathrm{e}$ (≈ 4.62).

approaching a straight line with slope 2, as in the previous case. The initial period during which R^* remains zero may correspond to the initial establishment phase. In (c), where $\gamma = 1$, the establishment period is even longer. Upon further investigation, we find that the establishment phase appears when γ is less than $4\pi/e(\approx 4.6)$ and that the smaller is the value of γ, the longer is this period. We will examine the length of the establishment period in more detail at the end of this section.

In section 2.5, we gave three possible reasons for the establishment phase. The one that applies here is the third reason: that is, the density of the initially invading individuals immediately falls below the threshold level due to diffusion, until the population increases through reproduction. The establishment phase is thus the period required for the population density to once again reach n^*.

During the expansion phase that follows, when R^* begins to increase monotonically from zero, R^* appears to lie on a straight line with slope 2 regardless of the γ value. To check this, we define the expansion rate as $C = R^*/T$, which is written, by using eqn (3.21), as

$$C = \frac{R^*}{T} = 2\left(1 + \frac{1}{T}\log\frac{\gamma}{4\pi T}\right)^{1/2}. \tag{3.23}$$

The right-hand side tends to 2 as $T \to \infty$. This shows that, during the expansion period, R^* asymptotically approaches a straight line with slope 2.

We now revert back to the original variables r^* and t, using eqn (3.20). By carrying out a scale conversion in which the abscissa and ordinate in Fig. 3.4 are multiplied by $1/\varepsilon$ and $\sqrt{D/\varepsilon}$, respectively, we obtain the actual range-versus-time curve for r^* and t. Clearly, this curve will also have an establishment phase when γ is below 4.6, and during the expansion phase the range expands at an almost constant rate, showing the type 1 pattern of the invasion curve. The actual expansion rate c is obtained by simply multiplying the ratio of the ordinate and abscissa scale conversions, $\sqrt{\varepsilon D}$, to the non-dimensionalized expansion rate $C = 2$, as follows:

$$c = \lim_{t \to \infty} \frac{r^*}{t} = 2\sqrt{\varepsilon D}. \tag{3.24}$$

This is the ultimate speed with which the population front proceeds outward from the origin. It shows that expansion occurs due to the combined effect of growth and diffusion, and that without either, no expansion takes place. Note that the magnitude of c does not depend on either the threshold density n^* or initial population size n_0.

Finally we determine the length of the establishment phase. If we define the establishment period t_e as the time interval between $t = 0$ and the moment at which expansion commences, it is the time when r^* of eqn (3.19) changes from zero to a positive value. So substituting $r^* = 0$ and $t = t_e$ into eqn (3.19), we have

$$\gamma = 4\pi\varepsilon t_e \exp(-\varepsilon t_e), \tag{3.25}$$

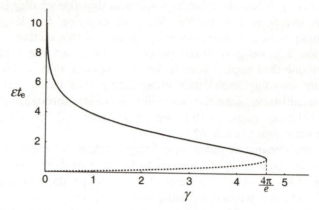

Fig. 3.5 Length of establishment phase as a function of γ. The lag period t_e rapidly decreases as γ increases from zero, and when γ exceeds $4\pi/e$, t_e abruptly becomes zero. Dotted line represents time taken for density of initially released individuals to fall below threshold density due to dispersal. (After Shigesada *et al.*, 1995)

where γ is given by eqn (3.22). If we plot eqn (3.25) using γ and εt_e for the abscissa and ordinate, respectively, we obtain Fig. 3.5. Thus, the establishment period t_e is infinitely large when $\gamma = 0$, but rapidly decreases with increasing γ, and abruptly disappears when $\gamma = 4\pi/e$ and above. Since $\gamma = \varepsilon n_0/Dn^*$, this implies that the establishment period t_e is prolonged as initial size n_0 and intrinsic growth rate ε decrease, or as diffusion coefficient D and threshold density n^* increase. If estimates of n, ε, t_e and D are available, the threshold density n^* can be established from eqn (3.25).

3.5 Fisher equation and travelling frontal wave

While Skellam's model can explain the spread pattern of type 1, it has the problem that the density within the population's range eventually becomes infinitely large, as seen in Fig. 3.3. To deal with this problem, we must resort to the Fisher equation (3.1):

$$\frac{\partial n}{\partial t} = D\left(\frac{\partial^2 n}{\partial x^2} + \frac{\partial^2 n}{\partial y^2} \right) + (\varepsilon - \mu n)n,$$

in which growth satisfies the logistic equation.

Since logistic growth contains a non-linear term, the Fisher equation cannot be solved explicitly as was the Skellam equation. So we begin by looking at how a numerically derived solution changes with time, to grasp the qualitative properties of a spreading population.

To this end, we first non-dimensionalize eqn (3.1) by using the following rescaling variables:

$$N(t) = \frac{n(t)}{K}, \quad X = \sqrt{\frac{\varepsilon}{D}}\, x, \quad Y = \sqrt{\frac{\varepsilon}{D}}\, y, \quad T = \varepsilon t, \qquad (3.26)$$

where $K = \varepsilon/\mu$, $N(t)$ denotes the non-dimensionalized density, and X, Y and T are respectively the dimensionless coordinates and time, similar to those used to make the scale conversion of expansion curve (3.19) into eqn (3.21). Equation (3.1) then becomes

$$\frac{\partial N}{\partial T} = \frac{\partial^2 N}{\partial X^2} + \frac{\partial^2 N}{\partial Y^2} + (1 - N)N. \qquad (3.27)$$

Thus the non-dimensionalized version of the Fisher equation contains no parameters, although the initial population size, $N_0 = n_0/K$, varies depending on the carrying capacity. In Fig. 3.6, we show the temporal change of the solution of eqn (3.27) numerically calculated. To compare this solution with that for the Skellam equation (3.16), we also non-dimensionalize eqn (3.16) by using the same rescaling variables of eqn (3.26), to obtain

$$\frac{\partial N}{\partial T} = \frac{\partial^2 N}{\partial X^2} + \frac{\partial^2 N}{\partial Y^2} + N. \qquad (3.28)$$

Note that this equation is equivalent to eqn (3.16) for the case $D = \varepsilon = 1$. Therefore, the solution of eqn (3.28) is also represented by Fig. 3.3 as such. Thus a comparison of Figs. 3.6 and 3.3 would clarify how the density effect of logistic growth influences the spatial pattern of expansion.

Let us first compare Figs. 3.6(a) and 3.3(a), which show the changes in spatial distributions for some period following the initial invasion. We see that the distributions display quite similar behaviour in both cases. Thus, we may say that the density of the initial propagules invading the origin rapidly decrease due to diffusion, during which time there is very little effect of competition. On the other hand, in contrast to Fig. 3.3(b), Fig. 3.6(b) shows that the population gradually recovers by reproduction, and consequently competition begins to manifest its effect, with the result that the density within the range approaches unity (the carrying capacity K in actual dimensions). The range front, meanwhile, forms a sigmoidal pattern and spreads at a constant speed. Such a solution, which advances at a constant speed while maintaining a constant shape of distribution, is called a 'travelling frontal wave'. As will be shown in section 3.9, the advancing speed of this wave is given by $c = 2\sqrt{\varepsilon D}$, which is identical to that of the

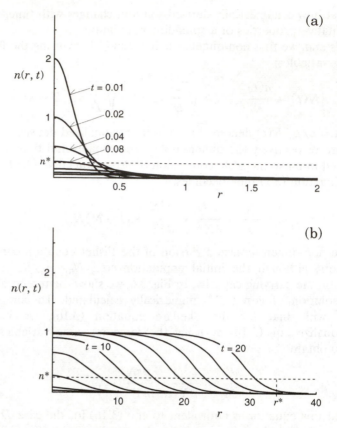

Fig. 3.6 Solution of Fisher model, with parameters $\varepsilon = 1$, $D = 1$, $\mu = 1$. (a) Change in distribution for $t = 0–1$. Spatio-temporal pattern resembles those seen in Figs. 3.1 and 3.3(a). (b) Change in distribution for $t = 2–20$. The front of the distribution tends to advance at constant velocity while maintaining its shape. Shape and location of leading edge are very similar to those from the Skellam equation (see Fig. 3.3b).

Skellam equation. Indeed comparing Figs. 3.3(b) and 3.6(b), we find that the shapes and locations of the leading edges are very similar. This is because in the vicinity of the range front, where the population density is always low and the effect of competition is therefore minimal, the equations for the two cases converge to the same form.

Combining these results together, we can say that as long as the threshold density, $N^* = n^*/K$, is smaller than 1 (i.e., the threshold density in actual scale n^* is less than the carrying capacity K), the establishment period t_e is very close to that determined from eqn (3.25).

While the above results are derived from numerical computation, it has been shown mathematically that such a travelling frontal wave indeed exists as the solution of eqn (3.1) and that it is stable against perturbations (Bramson, 1973). The derivation of the travelling wave and its speed as given by $c = 2\sqrt{\varepsilon D}$ is presented briefly in section 3.9. In summary, the relation between range distance and time given by the Fisher equation (3.1) has an initial establishment period when $\gamma \ (= \varepsilon n_0 / Dn^*) < 4\pi/e$, followed by an expansion phase in which the range expands at a fairly constant rate of $c = 2\sqrt{\varepsilon D}$, thus displaying the type 1 pattern of Fig. 2.12.

3.6 Travelling waves in general models

If the animal species under consideration has a tendency to orient itself toward external stimuli, or is carried by wind or water flow, the diffusion equation should be modified to incorporate drift or advection terms. For example, suppose that some insect expands its range by diffusion and Malthusian reproduction, and is also transported systematically by wind in a certain direction at a speed u. Then the change in the population density is described by the following equation (see section 3.8 for derivation):

$$\frac{\partial n}{\partial t} = D\left(\frac{\partial^2 n}{\partial x^2} + \frac{\partial^2 n}{\partial y^2}\right) - u\frac{\partial n}{\partial x} + \varepsilon n, \tag{3.29}$$

where the x-axis is aligned to the direction of the wind. With initial condition (3.3), the solution of eqn (3.29) is given by

$$n(x, y, t) = \frac{n_0}{4\pi Dt} \exp\left(\varepsilon t - \frac{(x - ut)^2 + y^2}{4Dt}\right).$$

This has the same form as eqn (3.4), except that x is replaced by $x - ut$. As shown in Fig. 3.7, for any fixed time t, contours of equal population density show concentric circles whose radius expands with time at a rate $2\sqrt{\varepsilon D}$, with its centre moving at a velocity u along the x-axis. Therefore, the speed of the front reaches a maximum, $2\sqrt{\varepsilon D} + u$, along the x-axis in the positive direction, and a minimum $2\sqrt{\varepsilon D} - u$ in the opposite direction.

When dispersal in two-dimensional space is not isotropic, so that the diffusion rates are different in various directions, we must deal with a more generalized diffusion equation. Van den Bosch *et al.* (1990) presented an explicit solution for the non-isotropic diffusion equation with advection terms. (See their seminal work for more details.) Wind-borne dispersal of seeds and pollen generally varies depending on multiple factors such as the settling velocity, height of release, wind speed and turbulence. Okubo and Levin (1989) have presented a sophisticated model incorporating these factors.

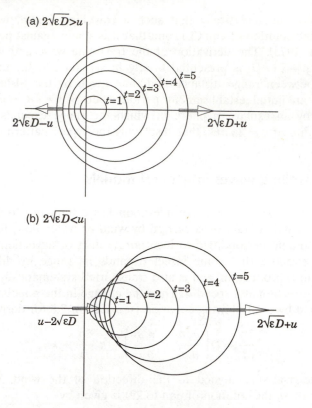

Fig. 3.7 Contours of equal density for the solution of the Skellam equation with advection term (3.29); u is the advection velocity and $2\sqrt{\varepsilon D}$ is the speed of the travelling wave when $u = 0$. (a) $u < 2\sqrt{\varepsilon D}$; (b) $u > 2\sqrt{\varepsilon D}$.

Finally we briefly discuss more general diffusion-reaction equations, in which either the growth rate or the diffusion coefficient is given by the general class of functions,

$$\frac{\partial n}{\partial t} = D\,\frac{\partial^2 n}{\partial x^2} + f(n)n, \tag{3.30}$$

where $f(n)$ is the per capita growth rate. When the growth function satisfies $f(n) > 0$ for $0 \le n \le K$, $f(K) = 0$, and $\mathrm{d}f/\mathrm{d}n \le 0$ for all $n \ge 0$ (the Fisher equation satisfies these conditions, with $f(0)$ corresponding to the intrinsic growth rate ε, and K being the carrying capacity), D is constant, and the initial distribution $n(x,0)$ is restricted in a finite domain instead of a pointwise distribution, it has been proved that the solution of eqn (3.30) converges asymptotically to a travelling wave like Fig. 3.6 and its asymptotic speed is given by $c = 2\sqrt{f(0)D}$ (Kolmogorov *et al.*, 1937; Aronson and Weinberger, 1975, 1978; Fife, 1979; Bramson, 1973). This suggests that density dependence of the growth rate is insignificant near the leading edge of the frontal wave if the growth rate $f(n)$ is positive at low density, i.e. $f(0) > 0$. From this, we may say that the Skellam or Fisher model, although elementary, will provide reasonable estimates for the rate of range expansion in terms of the pertinent parameters (Okubo, 1988; Levin, 1989). On the other hand, when the growth function $f(n)$ includes Allee effects, where $f(n)$ is negative at low densities, Lewis and Kareiva (1993) have shown that an invasion cannot propagate unless it initially exceeds a critical area, particularly for two-dimensional spread. This effect could prolong the period of the establishment phase as discussed in section 2.5.

As the population density increases, in some animals the population pressure tends to enhance dispersal, which is associated with interference between individual animals (Morisita, 1971; Shigesada *et al.*, 1979, 1980). For the particular case in which the diffusion coefficient increases with density, $D = dn$, and the growth function is logistic, Aronson (1980) and Newman (1980) independently proved that there exists a travelling wave solution and that its frontal speed is given by $c = \sqrt{\varepsilon d/2}$.

Diffusion models are often criticized because they assume that animals make completely random movements. Holmes (1993) showed that if the direction of random walk is correlated between consecutive jumps, the spatial distribution of organisms satisfies a telegraph equation (for a correlated random walk, see Kareiva and Shigesada, 1983). The solution of the telegraph equation combined with local population dynamics also exhibits a travelling wave, which has almost identical patterns as for the Fisher equation except that the tail is truncated at a finite distance. Holmes pointed out that the difference in wave speeds predicted from a telegraph equation and from the Fisher equation is less than 5% in various organisms.

Numerous authors have analysed eqn (3.30) for various other functional forms of $D(n)$ and $f(n)$ and initial distributions. Readers interested in this wide field of diffusion-reproduction processes should consult the works of Okubo (1980), Murray (1989), Renshaw (1991) and Banks (1994).

3.7 Match between theory and observation

As introduced in section 2.1, Skellam used the spread map of muskrats to derive the $\sqrt{\text{area}}$ -versus-time curve (Fig. 2.1(a, b)), and found that it fitted well with a straight line; then he proved that this property could be explained theoretically from eqn (3.16), based on Malthusian growth. Although this is where Skellam's investigation ended, if the diffusion coefficient D and intrinsic growth rate ε can be determined from microscale data on the demography and movements of individual organisms, it is possible to estimate the rate of spread from the formula $c = 2\sqrt{\varepsilon D}$. If this value agrees with the actual rate of spread as observed on a geographic scale, then it will further prove the validity of the Skellam equation as a framework for understanding the observed process of spread. Also, if D and ε are available for an organism which has just begun its invasion process, the rate of its future spread can be estimated.

In addition to Czechoslovakia, the muskrat has extended its range independently in regions within France, Denmark and Finland, and in each region its spread maps and microscale ecology have been surveyed and documented in detail. Williamson and Brown (1986) calculated the intrinsic growth rate of muskrats from a Leslie matrix (cf. Caswell, 1989):

$$\begin{pmatrix} 2.5 & 6.0 \\ 0.8 & 0.9 \end{pmatrix},$$

where muskrats are classified into first-year and older ones; the survival rates of first-year and older muskrats are 0.8 and 0.9; the mean number of female offspring per year from first-year and older ones are 2.5 and 6, respectively. The eigenvalue of the Leslie matrix which corresponds to e^{ε} is 4.03, thereby giving $\varepsilon = 1.39$/year. Independently, Andow *et al.* (1990) also estimated the intrinsic growth rate as $\varepsilon = 0.2$–1.1/year, which agrees fairly well with that by Williamson described above. In addition to ε, Andow *et al.* estimated D by using mark-recapture data on the basis of eqn (3.8) or (3.9); the results were $D = 51 \text{ km}^2/\text{year}$ and $D = 230 \text{ km}^2/\text{year}$ in the Netherlands and Finland, respectively. Substituting these values into $c = 2\sqrt{\varepsilon D}$, the range of the theoretically estimated speed was found to be $c = 6$–32 km/year. Meanwhile, they calculated the spread rate from maps of Czechoslovakia, France and Finland by using 'neighbourhood measurements' of spread (see section 2.6), and obtained the range of speed as $c = 1$–25 km/year. Thus the theoretical result agrees well with the observed one, although the former is slightly higher than the latter.

As we saw in section 2.1, Lubina and Levin (1988) analysed data on the range expansion of the California sea otter and found that the front of the range progressed at a constant speed within a homogeneous habitat: 1.4 km/year northward and 3.1 km/year southward from 1938 to 1972 (see Fig. 2.4). The otter's range centre drifted southward at a very low rate of 0.2–0.5 km/year, which is at least one order of magnitude smaller than the southerly flow of the offshore oceanic current. Thus they suggested that advection effects due to currents were not important in contributing to the observed differences between the northern and southern fronts. To test the applicability of the formula $c = 2\sqrt{\varepsilon D}$, Lubina and Levin made independent estimates of the diffusion coefficient and intrinsic growth rate based on data from the earlier years of the otter's invasion. As we discussed in section 3.5, for some period following initial invasion, the spatial distribution of the Skellam model, eqn (3.18), displays a pattern quite similar to that of the simple diffusion model, eqn (3.4). Therefore, if data on the spatial distribution at an early phase are available, the variance of the distribution should be $2Dt$ (for one-dimensional space), from which we can estimate the diffusion coefficient. On the other hand, for the earlier years in which non-linear density effects are negligible, the total population size should increase exponentially with exponent ε, from which the intrinsic growth rate could be evaluated. From these analyses, Lubina and Levin obtained $\varepsilon = 0.056$/year, and $D = 13.5$ km^2/year for the north and $D = 54.7$ km^2/year for the south. Substituting these parameters into $c = 2\sqrt{\varepsilon D}$ leads to the predicted front speeds, 1.74 km/year for the northern front and 3.50 km/year for the southern front. These values agree very well with the observed rate of spread of 1.4 km/year in the north and 3.1 km/year in the south between 1938 and 1972.

The range of the gypsy moth in North America has expanded since the turn of the century (Fig. 2.9). As mentioned in section 2.3, Liebhold *et al.* (1992) analysed historical records and determined that there were three periods of relatively constant rate of spread: about 9 km/year for 1900–16, 3 km/year for 1916–65 and 21 km/year for 1966–90. The population density along the leading edge of the gypsy moth infestation typically increases by a factor of 10–1000 from year to year (Elkinton and Liebhold, 1990). Thus if the population grows exponentially, the intrinsic growth rate is given as $\varepsilon = \ln\{$ratio of densities from year to year$\}$, which leads to $\varepsilon = 4.6$–6.9 if the ratio ranges from 100 to 1000 (Liebhold *et al.*, 1992). Meanwhile, the diffusion constant is estimated from the point release experiment conducted by Mason and McManus (1981). After attaching 1,100,000 gypsy moth eggs to trees within an area of 10 m radius, and monitoring the numbers of first instars trapped in sticky cylinders at distances of 60, 120 and 180 m in all directions after 13 days (i.e., long enough to cover the dispersal period of first instars), they found 35.5 larvae per cylinder at 60 m, 13.5 larvae per cylinder at 120 m and 5.7 larvae per cylinder at 180 m. Using the method of least squares to determine from eqn (3.5) the diffusion coefficient value that fits best with these data, we obtain $D = 0.003$ km^2 per generation, where it is assumed that gypsy moths undergo one generation change per year. Substituting these data into the formula $c = 2\sqrt{\varepsilon D}$, we obtain the estimated rate of spread to be about 0.2–0.3 km/year. This estimate is considerably less than the rates of spread observed during all periods including 1916–65 when control management was highly active. Thus Liebhold and his colleagues concluded that humans are inadvertently moving gypsy moth life stage beyond the infested area at a greater rate than would occur naturally.

Besides the rice water weevil (see section 2.3), Andow *et al.* (1990) also studied the spread of the agricultural pests the small cabbage white butterfly (*Pieris rapae*) and cereal leaf beetle (*Oulema melanopus*). Range expansion of the small cabbage white butterfly in Quebec is complicated by the fact that multiple invasions occurred, involving separate introduc-

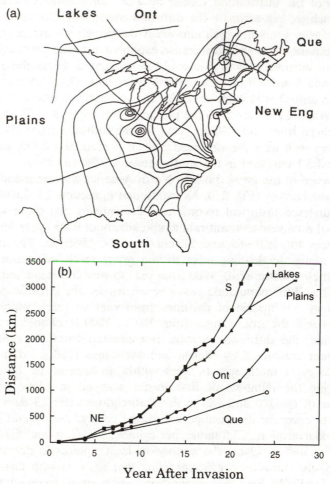

Fig. 3.8 Spread of small cabbage white butterfly in eastern North America (after Andow *et al.*, 1993). (a) Range expansion of small cabbage white butterfly. Map is divided into six sectors: Que, east through Maritime Provinces; Ont, west from Quebec City; Lakes, southwest along St Lawrence River and along Great Lakes; New Eng, south to Rhode Island; Plains, southwest along St Lawrence River and then west through Ohio and Great Plains; South, southwest through West Virginia, Alabama, Louisiana and Texas. (b) Range distance versus time for the six sectors.

tions as shown in Fig. 3.8(a). The map was examined by using the neighbourhood measurement (section 2.6), in which spread from Quebec is broken into six sectors. Figure 3.8(b) shows the range distance plotted against time for each sector. Overall, spread was non-linear, particularly in sectors towards the south; spread rates accelerated from an initial value of 28 km/year to 103 km/year in the Great Lakes and to 235 km/year toward the southwest. In Quebec, this insect completes three generations per year, whereas in Missouri, it completes six to seven generations per year. Thus, as it moved south, the population growth rate and dispersal rate would increase, accelerating the spread rate. From data on the daily movements of marked females (Jones *et al.*, 1980), Andow *et al.* estimated the mean displacements for butterflies to be between 0.5 and 1.2 km/day. Substituting these values for $\langle r \rangle$ in eqn (3.8) gives $D = 0.08$–0.46 km^2/day. To convert these daily rates into annual rates, the life expectancy for adults, 10–20 days, and the number of generations per year, 3–7, are multiplied, resulting in a yearly diffusion coefficient of $D = 2.4$–64 km^2/year. Meanwhile, the intrinsic growth rate was found to range between 9 and 31.5 per year. Substituting these parameter values into the formula $c = 2\sqrt{\varepsilon D}$ gives a predicted range of spread rate as 9.3–90 km/year, which is slightly lower than the observed range, 15–170 km/year.

Similarly, the invasion of cereal leaf beetle was analysed by Andow *et al.* to give $\varepsilon = 1.6$–1.9/year and $D = 0.4$ km^2/year, which leads to a predicted speed range of $c = 1.6$–1.7 km/year. This estimate is much smaller than

Table 3.1 Spread rates predicted from theory and those observed

Species	Intrinsic growth rate ε(year^{-1})	Diffusion coeff. D (km^2/year)	Spread rate c (km/year) Theor.	Obs.
Muskrat[1]	0.2–1.1	51–230	6–32	1–25
Cabbage butterfly[1]	9–32	2.4–64	9.3–90	15–170
Oulema melanopus[1]	1.6–1.9	0.4	1.6–1.7	27–90
Gypsy moth[2]	4.6	< 0.34	< 2.5	3–20
Sea otter[3]				
to north	0.056	13.5	1.74	1.4
to south	0.056	54.7	3.5	3.1
Black Death[4]	19	2.5×10^4	720	320–650
Rabies[5]	66	40–50	~ 70	30–60

[1] Andow *et al.* (1990)
[2] Liebhold *et al.* (1992)
[3] Lubina and Levin (1989)
[4] Nobel (1974)
[5] Yachi *et al.* (1989)

the actual spread rate, 27–90 km/year. This discrepancy is considered to be the result of some individuals travelling long distances via human transport systems and wind.

A comparison of the estimated spread rates using the Skellam model and observed rates for various organisms so far investigated by various authors is summarized in Table 3.1. The table also includes two additional cases, the invasions of rabies and bubonic plague, for which detailed accounts will be given in sections 9.4 and 10.3, respectively.

3.8 Appendix 1: derivation of diffusion equation

As stated in section 3.2, there are many ways to derive the diffusion equation. Here we use the random walk model (for a detailed treatment, consult Okubo, 1980, 1988; Hoppensteadt, 1982; and Murray, 1989).

Consider a one-dimensional line, as shown in Fig. 3.9, with numbered points spaced Δx apart. Each individual moves from a point to the adjacent one, with a probability of p to the right and q to the left. When $p > q$ ($p < q$), the random walk has a bias to the right (left); $p = q$ indicates an isotropic random walk. Let $n_{i,k}$ denote the population size at point i after k steps. Then the population at point i after the next step is the sum of those who remain without moving, $(1 - p - q)n_{i,k}$, and those arriving from the adjacent points, $qn_{i+1,k} + pn_{i-1,k}$. Thus we have

$$n_{i,k+1} = qn_{i+1,k} + (1 - p - q)n_{i,k} + pn_{i-1,k}. \qquad (3.31)$$

If we let $x = i\,\Delta x$ and $t = k\,\Delta t$, and assume that Δx and Δt are sufficiently small, we can rewrite eqn (3.31) as follows:

$$n(x, t + \Delta t) - n(x, t) = qn(x + \Delta x, t) - (p + q)n(x, t) + pn(x - \Delta x, t)$$
$$= \tfrac{1}{2}(p + q)\{n(x + \Delta x, t) - 2n(x, t) + n(x - \Delta x, t)\}$$
$$- \tfrac{1}{2}(p - q)\{n(x + \Delta x, t) - n(x - \Delta x, t)\}, \quad (3.32)$$

where $n(x, t)$ represents the population density at position x and time t.

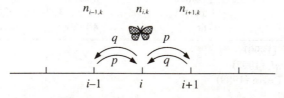

Fig. 3.9 Random walk model.

For both sides of eqn (3.32), we now carry out a Taylor expansion with respect to Δx and Δt about x and t, respectively, to obtain

$$\frac{\partial n(x,t)}{\partial t}\Delta t + O((\Delta t)^2) = \tfrac{1}{2}(p+q)\frac{\partial^2 n(x,t)}{\partial x^2}(\Delta x)^2$$

$$-\tfrac{1}{2}(p-q)\frac{\partial n(x,t)}{\partial x}(\Delta x) + O((\Delta x)^3).$$

If we take the limit for $\Delta x \to 0$ and $\Delta t \to 0$, and furthermore assume that $(p+q)\Delta x^2$ and $(p-q)\Delta x$ decrease so as to be of the same order of magnitude as Δt, that is, if

$$\lim_{\Delta x, \Delta t \to 0} \frac{(p+q)(\Delta x)^2}{2\,\Delta t} \to D,$$

$$\lim_{\Delta x, \Delta t \to 0} \frac{(p-q)(\Delta x)}{2\,\Delta t} \to u,$$

exist, we obtain

$$\frac{\partial n}{\partial t} = D\frac{\partial^2 n}{\partial x^2} - u\frac{\partial n}{\partial x}. \tag{3.33}$$

D is the diffusion coefficient indicating the degree of random dispersal. The velocity u is referred to as the advection (or transport) velocity: individuals are convected or transported towards the right (left) at a constant speed u, if u is positive (negative). When the probabilities of movement to the right and left are equal in a random walk, that is, $p = q$, u becomes zero and eqn (3.33) is reduced to a simple diffusion equation.

Applying a similar argument to the random walk on a two-dimensional lattice, we obtain the isotropic diffusion equation (3.2) or (3.29) with an advection term.

3.9 Appendix 2: travelling wave solution and its speed

In this section, we derive the travelling frontal wave which is the asymptotic solution of eqn (3.1). If the solution of eqn (3.1) is rotationally symmetric with respect to the origin, eqn (3.1) can be rewritten, using the radial distance r, as follows:

$$\frac{\partial}{\partial t}n = D\left(\frac{\partial^2 n}{\partial r^2} + \frac{1}{r}\frac{\partial n}{\partial r}\right) + (\varepsilon - \mu n)n. \tag{3.34}$$

Because the second term on the right-hand side, $(1/r)(\partial n/\partial r)$, approaches zero as the radius r increases, eqn (3.34) can be approximated by the following equation for sufficiently large r (Lewis and Kareiva, 1993):

$$\frac{\partial}{\partial t} n(r,t) = D \frac{\partial^2 n}{\partial r^2} + (\varepsilon - \mu n)n. \tag{3.35}$$

This means that, for large r, the distribution in the radial direction can be approximately described by eqn (3.35), which is equivalent to the one-dimensional Fisher equation. In subsequent discussions, we will refer frequently to diffusion taking place in one dimension and, as shown here, this can be considered as the spread taking place in the radial direction in two-dimensional space.

As we saw in section 3.5, a population satisfying the Fisher equation always evolves into a travelling frontal wave propagating at constant speed. Let us denote the solution of this travelling frontal wave by

$$n(r,t) = U(z), \quad z = r - ct, \tag{3.36}$$

where c is the wave speed. Equation (3.36) expresses the idea that $n(r,t)$, the spatial pattern of the travelling frontal wave, is moving to the right at speed c. Substituting this travelling wave form into eqn (3.35), we have

$$D \frac{d^2 U}{dz^2} + c \frac{dU}{dz} + (\varepsilon - \mu U)U = 0. \tag{3.37}$$

As we saw in Fig. 3.6, the density towards the rear of the travelling frontal wave has reached the carrying capacity ε/μ due to logistic growth, while the wave front approaches zero; hence

$$U(-\infty) = \varepsilon/\mu, \quad U(\infty) = 0. \tag{3.38}$$

If we can find c such that a non-negative solution $U(z)$ of eqn (3.37) that satisfies conditions (3.38) exists, $U(z)$ yields the travelling frontal wave with speed c. We shall see whether such a solution indeed exists, by converting eqn (3.37) to a set of first-order differential equations as follows:

$$\frac{dU}{dz} = V,$$

$$\frac{dV}{dz} = -\frac{c}{D} V - \frac{1}{D}(\varepsilon - \mu U)U. \tag{3.39}$$

Conditions (3.38) are then expressed as

$$U(-\infty) = \frac{\varepsilon}{\mu}, \quad V(-\infty) = 0,$$

$$U(\infty) = 0, \quad V(\infty) = 0. \tag{3.40}$$

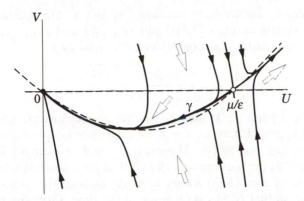

Fig. 3.10 Phase plane diagram for eqns (3.39). Dashed lines indicate null clines. Trajectory γ that connects $(0,0)$ and $(0, \varepsilon/\mu)$ represents the travelling wave solution.

Equations (3.39) have two equilibrium points, $(\varepsilon/\mu, 0)$ and $(0,0)$, which represent the respective (U, V) values at $z = -\infty$ and $z = \infty$, as given by eqns (3.40). Thus, if we can find a solution that connects the two equilibrium points and also satisfies $U(z) > 0$, then this is the travelling frontal wave's solution. Here we use the phase plane diagram to examine the trajectories for eqns (3.39) in the (U, V) plane (see Fig. 3.10). Let the plane be divided into regions depending on whether the right-hand sides of eqns (3.39) are either positive or negative. Then we have five regions separated by the parabola $cV = -(\varepsilon - \mu U)U$ and the U-axis, shown by broken lines. For each region, we indicate the sign for dU/dz and dV/dz, or the direction of change for (U, V) by a short arrow. Thus, the phase plane trajectory can be qualitatively drawn by following the arrows' direction, as shown in Fig. 3.10. In the vicinity of $(\varepsilon/\mu, 0)$ the trajectory forms a saddle. Therefore there are two trajectories emanating from $(\varepsilon/\mu, 0)$. If one connects to a trajectory leading to the origin while maintaining $U > 0$, as does curve γ in Fig. 3.10, this could be the travelling wave. To see whether γ approaches the origin while remaining entirely in the fourth quadrant, we shall examine the following equations, which are a linear approximation of eqns (3.39) in the vicinity of $(0,0)$:

$$\frac{dU}{dz} = V,$$

$$\frac{dV}{dz} = -\frac{c}{D}V - \frac{\varepsilon}{D}U. \tag{3.41}$$

The general solution of (3.41) is expressed as follows:

$$\begin{pmatrix} U \\ V \end{pmatrix} = a_1 e^{\lambda_1 z} \begin{pmatrix} c_{11} \\ c_{12} \end{pmatrix} + a_2 e^{\lambda_2 z} \begin{pmatrix} c_{21} \\ c_{22} \end{pmatrix}, \tag{3.42}$$

where a_1 and a_2 are arbitrary constants, λ_1 and λ_2 are eigenvalues of the coefficient matrix of eqns (3.41), and (c_{11}, c_{12}) and (c_{21}, c_{22}) are corresponding eigenvectors. The eigenvalues are given by

$$\lambda_{1,2} = \frac{-c \pm \sqrt{c^2 - 4\varepsilon D}}{2D}. \tag{3.43}$$

Since the real parts of λ_1 and λ_2 are always negative, U and V of eqn (3.42) both approach zero as $z \to \infty$. In other words, the trajectory invariably approaches the origin. However, λ_1 and λ_2 cannot be complex numbers, due to the requirement $U(z) > 0$. If they were complex numbers, the solution of eqn (3.42) would oscillate around the origin while approaching it, so that $U(z) < 0$ for some of the time. From the requirement that λ_1 and λ_2 be real numbers, we have

$$c \geq 2\sqrt{\varepsilon D}.$$

This is the condition that the velocity of the travelling frontal wave must satisfy. Conversely, it has been proved that whenever this condition is satisfied, there always exist a trajectory which connects the two equilibrium points while maintaining $U > 0$. While this implies that there are an infinite number of travelling frontal waves with velocities exceeding $2\sqrt{\varepsilon D}$, it has been proved mathematically that, if the initial distribution is concentrated in a localized region, the distribution will ultimately converge to a travelling frontal wave propagating at the minimum speed $c = 2\sqrt{\varepsilon D}$, and that this solution is a stable one which restores itself against local perturbations (Bramson, 1973; Fife, 1979).

4

Travelling waves in heterogeneous environments

4.1 Patch models

We have so far dealt with range expansions in homogeneous environments. As we saw in Chapter 2, however, the actual spreading range front is more or less irregularly shaped, never realizing concentric circles that spread at constant intervals as predicted by the Fisher model. Needless to say, the reason for this is that the invading species encounters rivers and hills in the course of its range expansion, and thus changes its rate of spread depending on the type of terrain. In this chapter, we study range expansion in a heterogeneous environment in which favourable and unfavourable habitats are distributed in patchwork fashion. When the invading species enters an unfavourable patch, it will be able to expand its range only if it successfully survives in that patch and reaches a favourable one lying ahead. If unfavourable patches dominate, the population may become extinct without expanding the range.

Habitat heterogeneity could affect spread via either habitat-dependent rates of movement or habitat-dependent rates of population increase. Here, we modify Fisher's model by assuming that the intrinsic growth rate ε and diffusion coefficient D vary with patches. In the following, we consider a situation in which two kinds of patch are arranged alternatively in one-dimensional space (see Fig. 4.1a), and organisms undergo dispersal as well as logistic growth as described by the following generalized Fisher equation:

$$\frac{\partial}{\partial t} n(x,t) = \frac{\partial}{\partial x} \left(D(x) \frac{\partial}{\partial x} n \right) + (\varepsilon(x) - \mu n)n, \qquad (4.1)$$

(a)

(b)

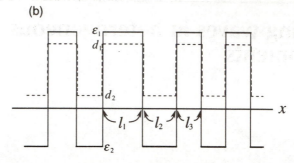

Fig. 4.1 (a) Patchy environment. Two kinds of patches, favourable and less favourable, are arranged alternately in one-dimensional space. (b) Diffusion coefficient $D(x)$ (dashed) and intrinsic growth rate $\varepsilon(x)$ (solid) as functions of x.

where the diffusion coefficient $D(x)$ and intrinsic growth rate $\varepsilon(x)$ are given by periodic step functions shown in Fig. 4.1(b). Thus,

$$D(x) = \begin{cases} d_1 \\ d_2 \end{cases}, \qquad \varepsilon(x) = \begin{cases} \varepsilon_1 & (\text{for } x_{2m} \leq x < x_{2m+1}) \\ \varepsilon_2 & (\text{for } x_{2m+1} \leq x < x_{2m+2}) \end{cases}, \qquad (4.2)$$

$$m = 0, \pm 1, \pm 2, \ldots,$$

where $x_0 = 0$, $x_{i+1} = x_i + l_{i+1}$ $(i = 0, \pm 1, \pm 2, \ldots)$, and l_i is the width of the ith patch. Odd-numbered patches are favourable ones with intrinsic growth rate ε_1 and diffusion coefficient d_1, whereas even-numbered patches are unfavourable ones with growth rate ε_2 and diffusion coefficient d_2. The intrinsic growth rate of unfavourable patches, ε_2, is lower than that of favourable patches, ε_1, $(\varepsilon_1 > \varepsilon_2)$, and can be negative (i.e., birth rate is lower than death rate). The relation between d_1 and d_2 varies with species so that no particular condition is imposed on them here. For instance, $d_1 < d_2$ for a species that speeds up on entering an unfavourable patch (such as the sea otter in section 2.1), while $d_1 > d_2$ if unfavourable patches hinder the species' movement. Clearly, in the case of $\varepsilon_1 = \varepsilon_2$ and $d_1 = d_2$, $d(x)$ and $\varepsilon(x)$ are constants and eqn (4.1) becomes the Fisher equation (3.1) for homogeneous environments.

By selecting appropriate values for the parameters ε_1, ε_2, d_1, d_2, and l_i $(i = 0, \pm 1, \pm 2, \ldots)$, it is possible to set up environments consisting of a mosaic of patches. However, since it is extremely difficult to obtain general solutions of eqn (4.1) when the patch width l_i is assumed to have an arbitrary value, we limit our attention to the following two special cases that are tractable mathematically: (1) a periodically varying environment in which two types of patches are regularly arranged, and (2) an irregularly varying environment in which the widths of favourable and unfavourable patches are assigned random values. We first deal with periodically varying environments, and then proceed to the case of irregularly varying environments.

4.2 Travelling periodic wave

Consider an environment which consists of a regular mosaic of patches, where favourable and unfavourable patches, with respective widths of l_1 and l_2, are arranged alternately. Namely we assume that

$$l_{2m+1} = l_1, \qquad l_{2m+2} = l_2 \quad (m = 0, \pm 1, \pm 2, \ldots).$$

When a propagule of some species is initially introduced to a locally confined region of the environment, under what circumstances does the species succeed in invading and expanding its range? Before undertaking a mathematical analysis, we begin by presenting the results of computer simulations of eqn (4.1) under varying sets of parameters. We found that the final destiny of the invading species ended in either of the two forms below, depending on the parameter values.

(I) The population, after dispersing somewhat from its invasion point, eventually becomes extinct (see Fig. 4.2(a).)

(II) The population evolves into a distribution propagating in both directions as shown in Fig. 4.2(b). The organisms increase rapidly in favourable patches and slowly in unfavourable ones so that the leading edge of the distribution fluctuates periodically without keeping a constant shape.

In addition to these two, one may consider a priori a situation where spreading ceases but a localized distribution range remains. However, no such case has been found in our simulations.

If we examine the propagating wave for case (II), we notice that any two wave shapes taken at a certain interval t^* can be perfectly superimposed by shifting the earlier one by the distance $l^* = l_1 + l_2$ (i.e., the spatial period). In other words, the following relation holds for the wave front advancing in the positive direction:

$$n(x, t - t^*) = n(x + l^*, t).$$

This type of wave, which changes shape periodically while continuously

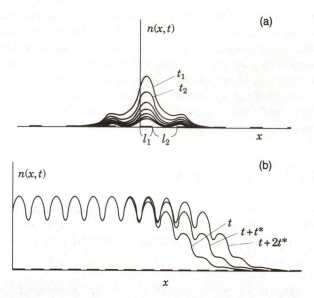

Fig. 4.2 Solutions of eqn (4.1) for periodic environment. (a) Localized initial population goes to extinction; $d_1 = 1$, $d_2 = 0.5$, $\varepsilon_1 = 1$, $\varepsilon_2 = -2$, $l_1 = 1$, and $l_2 = 2$. (b) Localized initial population evolves into travelling periodic wave. Travelling periodic waves on positive half plane at time t, $t + t^*$ and $t + 2t^*$ are plotted; $d_1 = 1$, $d_2 = 0.5$, $\varepsilon_1 = 1$, $\varepsilon_2 = -0.5$, $l_1 = 2$ and $l_2 = 1$. (After Shigesada *et al.*, 1986.)

expanding its range, is called a 'travelling periodic wave'. While the advancing speed of the travelling periodic wave will differ depending on whether the front is in a favourable or unfavourable patch, the mean velocity is given by the constant $c = l^*/t^*$, since the range distribution moves by $l^* = l_1 + l_2$ during time t^*. Thus, on the whole, the range-versus-time curve falls under type 1 (see Section 2.5), where range expansion takes place at a constant rate.

4.3 Condition for successful invasion

We now proceed to examine mathematically what conditions are required for cases (I) and (II) noted in the previous section.

For the invasion to be successful as in case (II), the population must increase when rare (if a small population is further reduced, it will disappear altogether). Expressed mathematically, this means that the zero solution $n(x) = 0$ of eqn (4.1), indicating the absence of the species in all areas, is unstable ($n(x) = 0$ is called stable if, for any $\varepsilon > 0$, there exists $\delta > 0$ such that if $\|n(x, 0)\| < \delta$, then $\|n(x, t)\| < \varepsilon$ for all $t > 0$; $n(x) = 0$ is unstable if it is not stable; see Britton (1986) for a more detailed definition). Conversely, case (I) corresponds to $n(x) = 0$ being stable, since any attempted initial introduction will end up in extinction.

In addition to the zero solution $n(x) = 0$, eqn (4.1) may have another equilibrium solution $n^e(x)$ which is positive and periodic with the spatial period $l^* = l_1 + l_2$:

$$\frac{\partial}{\partial x}\left(D(x)\frac{\partial}{\partial x}n^e\right) + (\varepsilon(x) - \mu n^e)n^e = 0,$$

$$n^e(x) = n^e(x + l^*).\tag{4.3}$$

We may imagine intuitively that if the zero solution $n(x) = 0$ is dynamically unstable while $n^e(x)$ is stable, then the transition from $n = 0$ to $n^e(x)$ may take place in a spatially mediated way. That is, a few organisms introduced into a local area will grow into an invasive propagating wave.

The stability of $n(x, t) = 0$ can be investigated by linearizing eqn (4.1) in the vicinity of $n = 0$ and solving the so-called eigenvalue problem. Thus, we obtain the condition under which $n(x, t) = 0$ is unstable, that is, the condition whereby a travelling periodic wave is generated, as follows (see derivation in section 4.7):

$$\tan\frac{L_1}{2} > \sqrt{-E_2 D_2}\,\tanh\left(\sqrt{-\frac{E_2}{D_2}}\,\frac{L_2}{2}\right),\tag{4.4}$$

where we denote

$$E_2 = \frac{\varepsilon_2}{\varepsilon_1}, \quad D_2 = \frac{d_2}{d_1}, \quad L_1 = \sqrt{\frac{\varepsilon_1}{d_1}}\,l_1, \quad L_2 = \sqrt{\frac{\varepsilon_1}{d_1}}\,l_2.\tag{4.5}$$

E_2 and D_2 are the intrinsic growth rate and diffusion coefficient, respectively, of the unfavourable patch relative to the favourable patch, while L_1 and L_2 are the widths of the favourable and unfavourable patches, respectively, using $\sqrt{d_1/\varepsilon_1}$ as the unit width. Conversely, if these four parameters do not satisfy eqn (4.4), the species will eventually become extinct so that the invasion fails. Figure 4.3 shows the respective zones in which invasion succeeds or fails, based on eqn (4.4). Figure 4.3(a) illustrates the (E_2, L_2) parameter space divided into regions where invasion can or cannot take place, respectively shown by the hatched and unhatched regions. The boundary curve dividing the two regions rapidly rises with increasing E_2, reaching infinity at a threshold value of $E_2^* = -(\tan L_1/2)^2/D_2$. Thus, as long as E_2 exceeds E_2^*, no matter how large the width L_2 of the unfavourable patch is, invasion is possible and a travelling periodic wave is generated. Figure 4.3(b), on the other hand, shows a similar division in the (D_2, L_2) space. In this case, the width L_2

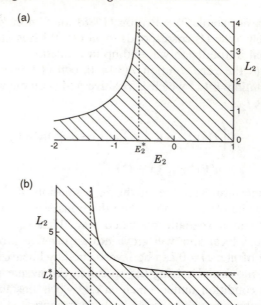

Fig. 4.3 Invasion conditions. Invasion succeeds in hatched regions where eqn (4.4) holds. Computer calculation demonstrates that localized initial distributions always evolve into a travelling periodic wave if the invasion is successful. (a) (E_2, L_2) plane for $D_2 = 0.5$ and $L_1 = 1$; (b) (D_2, L_2) plane for $E_2 = -0.5$ and $L_1 = 1$. (After Shigesada *et al.*, 1986.)

and diffusion coefficient D_2 of the unfavourable patch both have threshold values ($L_2^* = 2\tan(L_1/2)/(-E_2)$ and $D_2^* = (\tan L_1/2)^2/(-E_2)$, respectively), and if either parameter is less than its threshold value (i.e., $L_2 < L_2^*$ or $D_2 < D_2^*$), invasion will always succeed.

4.4 Speed of travelling periodic wave

We now determine the averaged speed $c = l^*/t^*$ of the travelling periodic wave's front. For convenience of discussion, instead of c, we use C_{rel}, defined as

$$C_{rel} = \frac{c}{2\sqrt{\varepsilon_1 d_1}}$$

As we saw in section 3.4, $2\sqrt{\varepsilon_1 d_1}$ corresponds to the spread rate in a homogeneous environment consisting only of a favourable patch (i.e., $l_2 = 0$), and so C_{rel} is the relative speed based on this. Although the method for determining the value of C_{rel} is basically the same as that used in section 3.9 for travelling frontal waves in homogeneous environments, a fairly tedious calculation is needed to reach the final result. Thus here we just give the resultant equation for the velocity of the travelling periodic wave (see Shigesada *et al.* (1986) for details):

$$C_{rel} = \frac{w^*}{2Y(w^*)} \qquad (4.6)$$

where

$$Y(w) = \frac{1}{L_1 + L_2} \log\left\{ G(w) + \sqrt{G(w)^2 - 1} \right\},$$

$$G(w) = \cosh q_1 L_1 \cosh q_2 L_2$$

$$+ \frac{q_1^2 + (D_2 q_2)^2}{2 D_2 q_1 q_2} \sinh q_1 L_1 \sinh q_2 L_2,$$

$$q_1 = \sqrt{w - 1}, \qquad q_2 = \sqrt{(w - E_2)/D_2},$$

and w^* is the solution of the following equation:

$$\frac{dY(w)}{dw} \frac{w}{Y(w)} = 1.$$

By numerically calculating eqn (4.6), we examined how the velocity C_{rel} varies with change in parameters E_2, D_2, L_1 and L_2.

Figure 4.4(a) shows the dependence of C_{rel} on E_2 for various L_2 values. C_{rel} decreases monotonically as E_2 decreases, becoming zero at a certain E_2 value. For smaller E_2, no travelling wave is generated.

Figure 4.4(b) shows the dependence of C_{rel} on D_2 for varying L_2. When the width of the unfavourable patch L_2 is smaller than the threshold value L_2^* (identical to L_2^* in Fig. 4.3(b)), C_{rel} increases monotonically with D_2. On the other hand, if L_2 is greater than L_2^*, C_{rel} shows a one humped curve which hits the D_2 axis at a certain value of D_2. The reader may consider this finding to be contrary to what intuition may suggest, reasoning that the proportion of individuals arriving at the favourable patch should increase if the diffusion coefficient of the unfavourable patch D_2 increases, thus causing the spread rate to increase monotonically as well. In actuality, however, if the diffusion coefficient in the unfavourable patches is high, this raises the rate at which individuals move from the favourable to the unfavourable patches; moreover, if the unfavourable patch stretches over a sufficiently large distance, many individuals die out before they reach the next favourable patch, thus lowering the spread rate.

Figure 4.4(c) shows the effect of the patch widths, L_1 and L_2, on C_{rel}. Each curve shows the change in C_{rel} with L_1 and L_2 for a fixed patch width ratio, $L_2/L_1 = l_2/l_1 = \nu$. Increasing the patch widths for a fixed

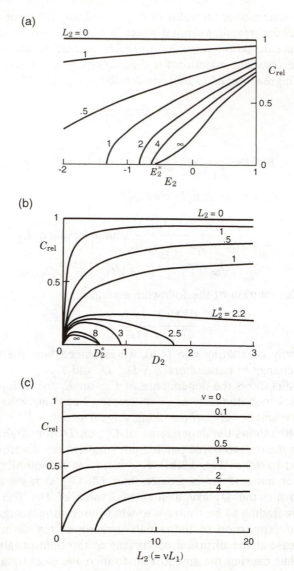

Fig. 4.4 Relative speed of travelling periodic wave C_{rel} as a function of parameters E_2, D_2 and L_2 ($= \nu L_1$): (a) C_{rel} versus E_2 for varying L_2; $D_2 = 0.5$, $L_1 = 1$; (b) C_{rel} versus D_2 for varying L_2; $E_2 = -0.5$, $L_1 = 1$; (c) C_{rel} versus L_2 for various fixed values of ν ($= L_2/L_1$); $D_2 = 0.5$, $E_2 = -0.5$. L_2^*, D_2^* and E_2^* correspond to asymptotes in Fig. 4.3. (After Shigesada *et al.*, 1986.)

ratio $L_1/L_2 = \nu$ means that the mosaic structure of the patches is enlarged in scale without changing the average environmental properties. As the patch widths are increased, all curves display a monotonic increase, eventually tending to a constant value. This tendency is exaggerated with increase in the value of ν. As seen from the curve for $\nu = 4$, no invasion takes place when the patch widths are small, but as the patch scale is expanded C_{rel} begins to take positive values, thus producing a travelling periodic wave.

While eqn (4.6) or Fig. 4.4 gives the exact relation between the relative propagation speed C_{rel} and its parameters, Shigesada *et al.* (1986) have also shown that the actual propagation speed c is approximately given by the following equation when the patch widths, l_1 and l_2, are sufficiently small:

$$c = \begin{cases} 2\sqrt{\langle \varepsilon \rangle_a \langle d \rangle_h} & \text{(for } \langle \varepsilon \rangle_a \geq 0) \\ 0 & \text{(for } \langle \varepsilon \rangle_a < 0) \end{cases}, \qquad (4.7)$$

where $\langle \varepsilon \rangle_a$ is the spatial arithmetic mean of $\varepsilon(x)$, and $\langle d \rangle_h$ the spatial harmonic mean of $d(x)$, respectively defined by

$$\langle \varepsilon \rangle_a = \lim_{l \to \infty} \frac{1}{2l} \int_{-l}^{l} \varepsilon(x)\,\mathrm{d}x = \frac{\varepsilon_1 l_1 + \varepsilon_2 l_2}{l_1 + l_2},$$

$$\langle d \rangle_h = \frac{1}{\displaystyle\lim_{l \to \infty} \frac{1}{2l} \int_{-l}^{l} \frac{1}{d(x)}\,\mathrm{d}x} = \frac{l_1 + l_2}{\dfrac{l_1}{d_1} + \dfrac{l_2}{d_2}} \qquad (4.8)$$

We can see that eqn (4.7) has the same form as $2\sqrt{\varepsilon_1 d_1}$, the spread rate in a homogeneous environment derived from the Fisher equation. In other words, for small patch widths l_1 and l_2, the speed of a travelling periodic wave is identical to that of a travelling frontal wave propagating in a homogeneous space, for which the intrinsic growth rate and diffusion coefficient are respectively given by $\langle \varepsilon \rangle_a$ and $\langle d \rangle_h$. It should be noted, however, that the harmonic mean gives a greater weight to the smaller diffusion coefficient (e.g., if d_1 is small enough to satisfy $l_1/d_1 \gg l_2/d_2$, $\langle d \rangle_h$ is approximately given by $d_1(l_1 + l_2)/l_1$, no matter how large d_2 is). Furthermore, as we saw in Fig. 4.4(c), even if $\langle \varepsilon \rangle_a$ is negative (i.e. when $\nu \geq 2$) so that the range cannot expand when the patch scale is small, a travelling periodic wave could begin to evolve when patch scale exceeds a certain level.

4.5   Travelling irregular wave

In this section, we consider invasion in an irregularly changing environment. As before, we use eqn (4.1) as the basic equation. However, the widths of the favourable and unfavourable patches, l_{2m+1} and l_{2m+2} ($m = 0, \pm 1, \pm 2, \ldots$), are respectively assigned random values uniformly chosen from the following intervals:

$$\bar{l}_1 - \sigma_1 < l_{2m+1} < \bar{l}_1 + \sigma_1,$$

$$\bar{l}_2 - \sigma_2 < l_{2m+2} < \bar{l}_2 + \sigma_2. \qquad (4.9)$$

$\bar{l}_1$ and $\bar{l}_2$ are the mean values of the widths of favourable and unfavourable patches, while $\sigma_1(\leq \bar{l}_1)$ and $\sigma_2(\leq \bar{l}_2)$ are the respective variations. Thus, the larger the σ_i's are, the more irregular the environment is, whereas if σ_1 and σ_2 are both zero, then the problem merely reverts to the case of a periodically varying environment as described in the previous section.

We carried out computer simulations of eqn (4.1) with the above conditions, and found as in the periodic changing environment that the solutions can be classified, depending on the parameter values, into the following two cases:

(I) The population will eventually become extinct so that invasion fails.
(II) The population evolves into a propagating wave as shown in Fig. 4.5. The range pattern changes irregularly in time and space, reflecting the random patch widths.

The propagating wave which changes shape irregularly while continuously expanding its range is called a 'travelling irregular wave'. We define the speed for such a wave as follows. As in section 3.2, we denote the threshold density for detection by n^*, and the range front at which $n(x,t)$ has just reached n^* by $x^*(t)$ (see Fig. 4.5). Then the advancing speed of

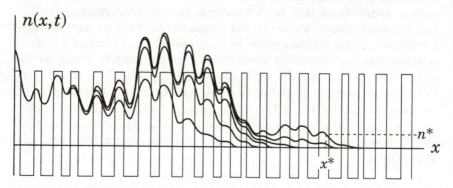

Fig. 4.5 Spatio-temporal changes in population density in an irregularly varying environment; $d_1 = 1$, $d_2 = 0.5$, $\varepsilon_1 = 1$, $\varepsilon_2 = -0.5$, $l_1 = 1$, $l_2 = 1.5$ and $\sigma_1 = \sigma_2 = 0.5$. Localized initial population evolves into a travelling irregular wave. Dashed line indicates threshold density, n^*; x^* represents range distance. (After Shigesada *et al.*, 1987.)

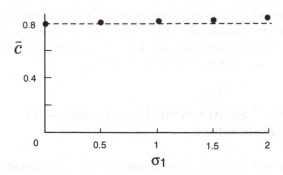

Fig. 4.6 Time-averaged speed $\bar{c}$ of travelling irregular waves as a function of σ_1. The ratio σ_1/σ_2 is fixed at 1; $d_1 = 1$, $d_2 = 0.5$, $\varepsilon_1 = 1$, $\varepsilon_2 = -0.5$, $l_1 = 2$ and $l_2 = 3$. Broken line indicates the speed of a travelling periodic wave in a corresponding periodic environment. (After Shigesada *et al.*, 1987.)

the front is given by dx^*/dt, which varies irregularly with time. Thus we introduce a time-averaged speed of dx^*/dt as:

$$\bar{c} = \lim_{t \to \infty} \frac{1}{t} \int_0^t \frac{dx^*(t')}{dt'} \, dt' = \lim_{t \to \infty} \frac{x^*(t)}{t}. \qquad (4.10)$$

If this limit converges, we may regard it as the mean speed of the travelling irregular wave. When $\bar{c}$ thus defined was calculated, we found that, for a given set of parameters, it invariably converged to a single value regardless of the threshold density n^* or initial values. Figure 4.6 shows how this $\bar{c}$ varies with σ_i. We can see that, even when the degree of irregularity, σ_i, is quite large, $\bar{c}$ remains close to the speed of the travelling periodic wave that develops in the corresponding periodically varying environment (where $\sigma_1 = \sigma_2 = 0$). Furthermore, it has been shown (Shigesada *et al.*, 1987) that if the patch widths, l_i, are sufficiently small, the mean speed $\bar{c}$ can be approximated by an equation similar to (4.7) in the previous section:

$$\bar{c} = \begin{cases} 2\sqrt{\langle \varepsilon \rangle_a \langle d \rangle_h} & (\text{for } \langle \varepsilon \rangle_a \geq 0) \\ 0 & (\text{for } \langle \varepsilon \rangle_a < 0) \end{cases},$$

where

$$\langle \varepsilon \rangle_a = \frac{\varepsilon_1 \bar{l}_1 + \varepsilon_2 \bar{l}_2}{\bar{l}_1 + \bar{l}_2}, \qquad \langle d \rangle_h = \frac{\bar{l}_1 + \bar{l}_2}{\dfrac{\bar{l}_1}{d_1} + \dfrac{\bar{l}_2}{d_2}}.$$

In summary, we can state that the mean velocity of a travelling frontal wave generated in an irregularly varying environment can be approximated by the velocity of a travelling periodic wave in the corresponding periodically varying environment (i.e., the values shown in Fig. 4.4). Thus, while the range-versus-time curve locally fluctuates, overall, it follows the type 1 pattern, where spreading takes place at a constant speed.

4.6 Effects of environmental fragmentation on biological conservation

So far, using eqn (4.1), we have examined the environmental conditions under which an alien species succeeds in invading a space and expands its range. In fact, the two problems, 'whether a species can invade a certain type of environment' and 'whether the species placed under that environment can ultimately persist', are equivalent in mathematical terms. Thus, in this section we shift our perspective and ask whether a species can continue surviving when its original habitat has been disturbed, thus creating areas unfavourable to its survival.

We can find many situations, often in the context of environmental issues, in which areas unfavourable to a species' presence are being created, as when forests are divided up by roads or cultivated fields thus segmenting the species' habitat; or when an isolated stand of forest is gradually whittled away from its perimeters.

What is the minimum area of favourable patches necessary for a species to survive? If a fixed area of favourable environment is allowed to remain, are there ways of dividing that area that are less harmful to that species? For instance, which is better from an environmental viewpoint: one large reserve or finely fragmented ones? We examine such issues by considering the following two idealized systems.

I. *Fragmented environment.* As shown in Fig. 4.7(a) the natural environment is segmented into belts, with favourable and unfavourable patches of widths l_1 and l_2, respectively, arranged alternately. The species in question has intrinsic growth rates ε_1 and ε_2 ($\varepsilon_1 > \varepsilon_2$) and diffusion coefficients d_1 and d_2 in the respective patches. This corresponds to the above-mentioned example where roads, running in parallel at constant intervals, cut through a favourable environment; the roads are the unfavourable patches, while the remaining areas are the favourable ones. Individuals located in the roads are deprived of their habitats and must generally endure adverse conditions, such as the high risk of being run over by vehicles, resulting in a high death rate (i.e., $\varepsilon_2 < 0$).

II. *Isolated environment.* As shown in Fig. 4.7(b), an isolated favourable patch of width l_1 is surrounded on both sides by unfavourable patches of width $l_2/2$. The intrinsic growth rates and diffusion coefficients in

(a)

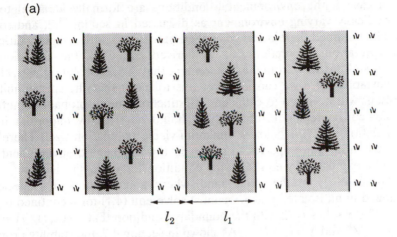

l_2 l_1

(b)

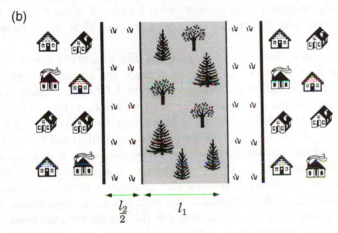

$\dfrac{l_2}{2}$ l_1

Fig. 4.7 Fragmented environment: (a) periodic segmented environment; (b) isolated environment (after Shigesada, 1992).

the two patch types are ε_1 and ε_2, and d_1 and d_2, respectively. Here any individual encountering the outer edge will turn back instead of crossing. In mathematical terms, this is equivalent to the case where the distribution at the outer perimeters satisfies the reflecting (Neumann) boundary condition, $\partial n(x,t)/\partial x = 0$. This case corresponds to a situation where the forest is surrounded by cultivated fields (or roads), outside which lie human settlements; the animals living in the forest may occasionally risk an excursion out to the fields but avoid entering areas of human habitation.

In case I, the environmental conditions are formally identical to the periodically varying environment as discussed in section 4.2, and so the results obtained therein can be applied here, albeit with a slightly different interpretation. In section 4.3, we derived eqn (4.4), the condition for successful invasion, from the requirement that the zero solution, $n(x, t) = 0$, is unstable. Here we can interpret this to mean that the species initially present will never go to extinction. (Extinction due to probabilistic effects is not considered here, although it should be crucially important if the population size falls below a minimum viable population size.) Therefore, the condition for a species to persist in the fragmented environment can also be expressed by eqn (4.4), the condition for successful invasion.

For case II, on the other hand, where a population of a species is trapped in an isolated patch, we must solve eqn (4.1) for a confined region $-l_2/2 \leq x \leq l_1 + l_2/2$ with the boundary condition that $\partial n(x, t)/\partial x = 0$ at $x = -l_2/2$ and $x = l_1 + l_2/2$. As shown in section 4.7, the stability property of the zero solution in the present case exactly conforms to that in the periodically changing environment given by case I. Thus it turns out that the condition for the species to survive in case II is given by eqn (4.4) as well.

Since we see that eqn (4.4) provides the condition for a species to survive in both cases, I and II, if we replace the concept of 'invasion' by 'survival', the discussion presented in section 4.3 can be directly applied to issues of environmental conservation. In Fig. 4.3, therefore, if the environmental parameters fall within the hatched area, the population will survive, while if they lie in the unhatched area, the population will become extinct. Focusing on the environment of case II, we illustrate in Fig. 4.8 how the spatial distribution of a population changes with increasing l_2 while the total width $l_1 + l_2$ is fixed. We find that as the size of the favourable patch is reduced (i.e., l_2 is increased), the population density, which shows a gradual decline from the centre towards the boundaries, decreases in both favourable and unfavourable patches, and when l_2 exceeds a certain value, the population becomes extinct throughout the entire region. These results provide a theoretical explanation of what Lawton (1995) pointed out in some documented cases of insects and mammals: their overall survival is sustained by supplementing sink populations in unfavourable patches with immigrants from source populations in favourable patches (Rogers and Randolph, 1986; Caughley et al., 1988), and hence the destruction of the source would make the sink populations unsustainable.

Turning our focus to case I, if the relative sizes of favourable and unfavourable patches are fixed, we can say from Fig. 4.4(c) that keeping patches in large blocks is better than fragmenting the environment into small ones (i.e., if l_2/l_1 is constant, the absolute values of l_1 and l_2 should be as large as possible) from the viewpoint of species preservation.

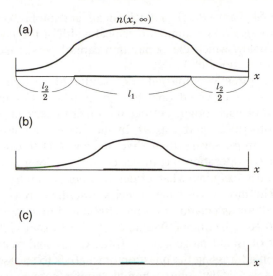

Fig. 4.8 Change in spatial distribution with increase in l_2, while total width is fixed as $l_1 + l_2 = 4$: (a) $l_2 = 2$; (b) $l_2 = 3$; (c) $l_2 = 3.6$. Other parameters are chosen as $d_1 = 1$, $d_2 = 0.5$, $\varepsilon_1 = 1$, $\varepsilon_2 = -0.5$ and $\mu = 1$. When l_2 is larger than 3.14, the population goes to extinction.

The conditions for survival discussed above were derived for a highly simplified system. Thus, these models would require modifications when we apply them to actual situations. Cantrell and Cosner (1991), for instance, have extended case II to deal with a favourable patch surrounded by two unfavourable patches of different widths on both sides. They found that, if the total width of the unfavourable patches is kept constant, the closer the favourable patch lies to one of the boundaries, the greater is the overall suitability of the environment.

Cantrell and Cosner (1991) further modified the above model by replacing the reflecting boundary condition with an absorbing one, that is, $n(x,t) = 0$ at both outer boundaries. The absorbing condition represents a situation where the environment is surrounded by a completely hostile exterior region so that organisms that reach the boundaries will perish and never return. The analysis of this model suggests that, in contrast with the case of the reflecting boundary condition, centring the favourable patch gives a more suitable environment than locating it near a boundary. The explanation for this is that unfavourable patches somewhat insulate the population from the hostile exterior if the favourable patch is far from the boundary (see also, Seno, 1989). Note here that when $l_2 = 0$ (i.e., the favourable patch is directly surrounded by a hostile exterior region), the above model is reduced to the pioneering model on critical patch size presented by Skellam (1951) and Kierstead and Slobodkin (1953). They showed that there exists a minimum critical width of the favourable patch, $\pi(d_1/\varepsilon_1)^{1/2}$, below which the population cannot persist (see comprehensive review by Okubo, 1980).

Because we directly employed the results of section 4.3 here, the environment was segmented only in one direction, but in reality roads usually crisscross each other, cutting the environment in both directions (e.g., along both the *x*- and *y*-axes). In such cases, the survival condition will clearly be more severe than that of eqn (4.4). Furthermore, if the species under consideration has direct or indirect interactions with other species, we must examine what effect fragmentation has on the entire community. The models for single species described above were extended to a prey–predator association by Seno (1991) and to competing species by Shigesada and Roughgarden (1982) and Ali and Cosner (1995).

The theory of island biogeography (MacArthur and Wilson, 1967) has been applied in addressing the question of whether to choose a single large reserve or finely divided ones when designing nature reserves for the purpose of maintaining species diversity. In this theory, the number of species found on an island represents a dynamic equilibrium between the rate of colonization by new species and the rate of extinction of existing species. Immediate results of the theory suggest that reserves should be large, rounded, and close to other reserves (Gilpin and Diamond, 1980; Diamond and May, 1981; Harrison, 1994). However, in recent years an increasing number of studies on habitat fragmentation, in empirical examples and in theoretical models, have provided different and often conflicting conclusions depending on the particular environments or on the extent of the spatial or temporal scales that govern the system in question (see Simberloff and Abele, 1976; Higgs and Usher, 1980; Soule and Wilcox, 1980; Kareiva, 1982, 1987; Wilcove *et al.*, 1986; Quinn and Hastings, 1987; Gilpin, 1988; Doak *et al.*, 1992; May, 1994a; Tilman *et al.*, 1994; Kareiva and Wennergren, 1995).

Although we have employed the diffusion-reaction model in this chapter, metapopulation models may be another promising approach to examine the effect of fragmentation on the persistence of populations. Metapopulations are made up of many subpopulations distributed among a mosaic of patches. In contrast to the diffusion-reaction model, each patch is taken as the basic unit; population size or spatial structures within patches are ignored (but see Hastings, 1991; Gyllenberg and Hanski, 1992). Lande (1987, 1988) analysed the effect of patch removal (habitat destruction) in the framework of metapopulation theory and demonstrated that there exists a threshold fraction of suitable patches that must be left for a single species to persist in a region. For the case of metapopulations consisting of prey and predators or competing species, Nee and May (1992) and May (1994a) emphasized that the effect of habitat destruction can bring about intuitively non-obvious changes in community structures, depending on the degree of habitat destruction (for example, patch removal could increase the biodiversity in a competition system). Readers interested in metapopulation in conservation are recommended to consult the book edited by Gilpin and Hanski (1991) and also reviews by Wilcove *et al.* (1986), Harrison (1994), Lawton *et al.* (1994) and Hastings (1994).

4.7 Appendix: derivation of invasion condition, eqn (4.4)

Assuming that $D(x)$ and $\varepsilon(x)$ of eqn (4.1) are given by periodic step functions (where the favourable and unfavourable patches have constant widths of l_1 and l_2, respectively), we determine the condition under which the equilibrium solution $n = 0$ is dynamically unstable (i.e., the condition for successful invasion). To this end, we first linearize eqn (4.1) about $n(x) = 0$ to obtain

$$\frac{\partial}{\partial t} n(x,t) = \frac{\partial}{\partial x}\left(D(x) \frac{\partial}{\partial x} n(x,t)\right) + \varepsilon(x)n(x,t). \qquad (4.11)$$

By substituting $n(x,t) = \varphi(x)\exp(\lambda t)$ into eqn (4.11), we have

$$\frac{d}{dx}\left(D(x) \frac{d}{dx} \varphi(x)\right) + (\varepsilon(x) - \lambda)\varphi(x) = 0, \qquad (4.12)$$

where $\varphi(x)$ is the characteristic function and λ is its eigenvalue. Solving eqn (4.12) for φ under the conditions that φ and $D(x)d\varphi/dx$ are continuous and periodic with period $l_1 + l_2$, we have the following equation (Shigesada *et al.*, 1986):

$$\sqrt{1 - \lambda/\varepsilon_1} \, \tan\left(\sqrt{1 - \lambda/\varepsilon_1} \, L_1/2\right)$$
$$= \sqrt{(-E_2 + \lambda/\varepsilon_1)D_2} \, \tanh\left\{\sqrt{(-E_2 + \lambda/\varepsilon_1)D_2} \, L_2/2\right\}, \quad (4.13)$$

where E_2, D_2, L_1 and L_2 are defined by eqn (4.5). The solutions of eqn (4.13) provide the eigenvalues of eqn (4.12).

It has been proved mathematically (Magnus and Winkler, 1966; Coddington and Levinson, 1972) that the trivial solution $n = 0$ of eqn (4.11) is unstable if $\lambda_0 > 0$, and stable if $\lambda_0 < 0$, where λ_0 is the maximum eigenvalue. This means that the stability property bifurcates from

$\lambda_0 = 0$. Thus, substituting $\lambda = 0$ into eqn (4.13), we obtain the stability condition as

$n = 0$ is unstable (i.e., $\lambda_0 > 0$) when

$$\tan(L_1/2) > \sqrt{-E_2 D_2}\ \tanh\left(\sqrt{-E_2/D_2}\ L_2/2\right),$$

$n = 0$ is stable (i.e., $\lambda_0 < 0$) when

$$\tan(L_1/2) < \sqrt{-E_2 D_2}\ \tanh\left(\sqrt{-E_2/D_2}\ L_2/2\right). \quad (4.14)$$

The upper relation leads to eqn (4.4).

When a population is confined in an environment which consists of a favourable patch surrounded by unfavourable ones on both sides as shown in Fig. 4.7(b), we must solve eqn (4.1) under boundary conditions $\partial n(x,t)/\partial x = 0$ at $x = -l_2/2$ and $x = l_1 + l_2/2$. The stability property of the trivial solution $n = 0$ can be derived by examining eqn (4.11) under boundary conditions $\partial n(x,t)/\partial x = 0$ at $x = -l_2/2$ and $x = l_1 + l_2/2$. It can be shown that the characteristic equation of that equation and its eigenvalues are given by the same equations as (4.12) and (4.13), respectively, and hence we have the same conclusion as eqns (4.14) for the present case.

5

Invasion by stratified diffusion

5.1 Short-distance dispersal and long-distance dispersal

So far, we have treated the movement of organisms as a diffusion process based on random walks. As we saw in Chapter 2, however, there are some species of birds, insects, plants and herding animals that expand their ranges not only by random movement into surrounding adjacent areas, but also by long-distance dispersal. When various modes of dispersal occur side by side within a species, Hengeveld (1989) proposed that such a dispersal process be called 'stratified diffusion' (but, in epidemiology, 'hierarchical diffusion' is in use as an alternative term).

A species spreading by stratified diffusion often shows an accelerating range-versus-time curve, as in type 2 or 3 of Fig. 2.12. In type 2 the spreading starts at a slow rate followed by linear expansion at a higher rate, while in type 3, the spread rate continually increases with time. As Andow *et al.* (1993) have pointed out, the initial speed of expansion could be mainly determined by neighbourhood diffusion of a founder population, while the accelerated expansion observed in the later phase is largely a reflection of the growth of new colonies successively created by long-distance migrants (see also Hengeveld, 1989).

As seen in section 2.5, in organisms that exhibit the type 3 pattern, the long-distance migrants tend to disperse far away from the parent population, so that their colonies expand independently of the parent population for a considerable period of time. In type 2, on the other hand, the long-distance migrants do not travel so far, and consequently their colonies are absorbed into the parent population's range in a relatively short time. How can we mathematically formulate these different situations on the basis of the combined effects of the short- and long-distance dispersal modes? In the following two sections, we introduce two models, the 'scattered colony model' and 'coalescing colony model' (Shigesada *et al.*, 1995), which respectively provide mathematical interpretations of the expansion patterns of type 3 and type 2. Applying the results of the analyses to some biological invasions reviewed in Chapter 2, we estimate parameter values that determine the underlying process of stratified diffusion. In section 5.5, more mechanistic models that incorporate dispersal distance distribution and/or age structure developed by Mollison (1972, 1977) and van den Bosch *et al.* (1990, 1992) will be described.

5.2 Scattered colony model

As shown in Fig. 5.1, we consider a situation in which the invading species extends its range into surrounding areas by random diffusion, while at the same time producing individuals that disperse far away. We assume that the colonies established by these long-distance migrants are sufficiently removed from each other so that each one remains an independent spreading range for a long period. The case of cheatgrass discussed in section 2.4 is of this type.

We begin at the outset ($t = 0$) when a few propagules invade a point and create there the nucleus of an isolated colony. As we saw in Chapter 3, if their main dispersal mode is random diffusion, the range spreads concentrically at a constant rate of c. If the environment is homogeneous, c is

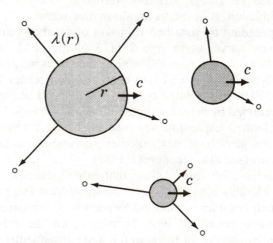

Fig. 5.1 Range expansion in the scattered colony model. The invading species expands its range at a constant rate c by neighbourhood diffusion and simultaneously produces long-distance dispersers which create nuclei of new colonization at rate $\lambda(r)$. Each nucleus expands at constant rate c and at the same time emits long-distance dispersers, as does its parent population. The nuclei are separated far enough from each other that their ranges independently expand for a long time. (After Shigesada *et al.*, 1995.)

given by $2\sqrt{\varepsilon D}$, where ε and D are the intrinsic growth rate and diffusion coefficient, respectively. If environmental conditions vary in patchwork fashion, the average rate of spread as derived in Chapter 4 is used for c. For sake of simplicity, moreover, we assume that the establishment period is negligibly short and so the range starts to expand at the constant rate c immediately after invasion. At the same time, this population will produce long-distance migrants who create nuclei of new colonies removed from the parent population. Let us denote by $\lambda(r)$ the number of nuclei produced per unit time by long-distance migrants dispersed from a parent colony of range radius r. Generally, successful colonization is achieved only when at least a few organisms (including a male–female pair in animals) settle in the same region and succeed in producing their offspring against the risk of extinction due to demographic stochasticity, inbreeding depression or any other causes (see section 2.5). Accordingly, $\lambda(r)$ represents the number of nuclei that overcome those various severe conditions. Just as in the parent population, each nucleus will expand its range area by short-distance dispersal (also referred to as 'neighbourhood diffusion') at a rate c, while simultaneously producing long-distance migrants. When this process is repeated over and over, the number of colonies will increase exponentially. How then will the total area of the colonies increase with time?

From among the colonies widely scattered about, we now denote by $\rho(r,t)\,dr$ the number of colonies that have a radius in the range r to $r + dr$ at time t; $\rho(r,t)$ is thus a density function with regard to the colony size r. Since the total invaded area, which we denote by $A(t)$, is the sum of these colonies, we can write

$$A(t) = \int_0^\infty \pi r^2 \rho(r,t)\,dr. \tag{5.1}$$

Thus the range expansion is determined from $\rho(r,t)$, which we derive in the following.

A colony of radius r will grow to a colony of radius $r + c\,dt$ after a period dt. Therefore, the following holds:

$$\rho(r,t) = \rho(r + c\,dt, t + dt).$$

By expanding the right-hand side with respect to dt, dividing the entire equation by dt, and letting $dt \to 0$, we obtain

$$\frac{\partial \rho}{\partial t} + \frac{\partial c\rho}{\partial r} = 0. \tag{5.2}$$

This equation, generally known as the von Foerster equation (Metz and Diekmann, 1986; Hallam and Levin, 1986), expresses the fact that the scattered colonies are each expanding at a rate c due to short-distance dispersal.

Meanwhile, the process by which long-distance migrants establish the nuclei of new colonies is described by the following equations:

$$cp(0,t) = \int_0^\infty \lambda(r)p(r,t)\,dr, \qquad (5.3a)$$

$$p(r,0) = \delta(r). \qquad (5.3b)$$

The right-hand side of eqn (5.3a) expresses the number of new nuclei created by long-distance migrants per unit time, and this is equal to the left-hand side, which is the number of colonies with zero radius appearing per unit time. Equation (5.3b) means that the propagules initially introduced at a point create a nucleus of radius 0 at time $t = 0$. We thus have a set of equations, (5.2) and (5.3a,b), for the size distribution function $p(r,t)$. But we still need the function of the colonization rate, $\lambda(r)$, to solve this.

The colonization rate $\lambda(r)$ should be given by a non-decreasing function of radius r. Here we consider the following three cases:

$$\text{(a)} \quad \lambda(r) = \lambda_0,$$

$$\text{(b)} \quad \lambda(r) = \lambda_1 r, \qquad (5.4)$$

$$\text{(c)} \quad \lambda(r) = \lambda_2 r^2.$$

For case (a), every colony produces long-distance migrants at a constant rate, irrespective of its size. In (b), the number of long-distance migrants produced per unit time is proportional to the circumference of the parent colony. This may occur if the long-distance dispersers are produced only at the leading edge (periphery) of each colony. In (c), the production rate of long-distance dispersers per unit area is constant anywhere in the colony, so that the total rate per colony is proportional to its area. For each case (a), (b) and (c), we can solve eqns (5.2)–(5.3) to obtain explicit functions of $p(r,t)$ for $r \leq ct$ as (see Iwata *et al.*, 1996):

$$\text{(a)} \quad p(r,t) = \frac{\lambda_0}{c}\, e^{\lambda_0(t-r/c)} + \delta(r-ct),$$

$$\text{(b)} \quad p(r,t) = \frac{\omega_1}{2c}\, (e^{\omega_1(t-r/c)} - e^{-\omega_1(t-r/c)}) + \delta(r-ct), \qquad (5.5)$$

$$\text{(c)} \quad p(r,t) = \frac{\omega_2}{3c}\, \{e^{\omega_2(t-r/c)} + ke^{\omega_2 k(t-r/c)} + k^2 e^{\omega_2 k^2(t-r/c)}\}$$

$$+ \delta(r-ct),$$

where $\omega_1 = (c\lambda_1)^{1/2}$, $\omega_2 = (2c^2\lambda_2)^{1/3}$ and $k = (-1 + \sqrt{3}\,i)/2$. Figure 5.2

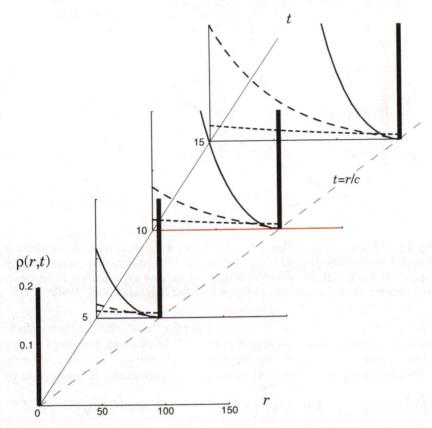

Fig. 5.2 Change of size distribution $\rho(r, t)$ with time: ----, for $\lambda(r) = \lambda_0$; ---, for $\lambda(r) = \lambda_1 r$; ——, for $\lambda(r) = \lambda_2 r^2$. Size of primary population increases along the thin dashed line; $\lambda_0 = 0.08$, $\lambda_1 = 0.004$, $\lambda_2 = 0.0004$ and $c = 10$.

illustrates how the size distribution changes with time. Substituting these results into eqn (5.1), we have the total area for the three cases of eqns (5.4a, b and c) respectively as

$$\text{(a)} \quad A(t) = \frac{2\pi c^2}{\lambda_0} \left\{ \frac{1}{\lambda_0} (e^{\lambda_0 t} - 1) - t \right\},$$

$$\text{(b)} \quad A(t) = \frac{\pi c^2}{\omega_1^2} (e^{\omega_1 t} + e^{-\omega_1 t} - 2), \qquad (5.6)$$

$$\text{(c)} \quad A(t) = \frac{2\pi c^2}{3\omega_2^2} \left\{ e^{\omega_2 t} + 2e^{-\omega_2 t/2} \sin\left(\frac{\sqrt{3}}{2} \omega_2 t - \frac{5}{6} \pi \right) \right\}.$$

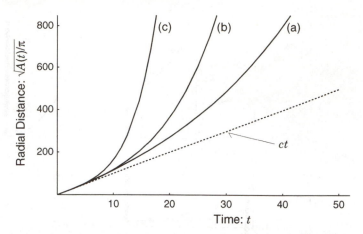

Fig. 5.3 Effective range radius $\sqrt{A(t)/\pi}$ for three different types of colonization rate: (a) $\lambda(r) = 0.08$; (b) $\lambda(r) = 0.004r$; (c) $\lambda(r) = 0.0004r^2$. The range initially expands at a constant rate c and then tends to increase exponentially in all cases, thus showing the expansion curve of type 3. (After Shigesada *et al.*, 1995.)

Taking the square root of $A(t)/\pi$, we obtain the effective range radius-versus-time curve, which is plotted in Fig. 5.3. In all cases, the range radius curve begins with slope ct and increases acceleratingly with time. After a sufficiently long time, they tend to increase exponentially in proportion to

$$\text{(a)} \quad \exp\{\lambda_0 t\}, \qquad \text{(b)} \quad \exp\{(c\lambda_1)^{1/2}t\}, \qquad \text{(c)} \quad \exp\{(2c^2\lambda_1)^{1/3}t\}, \quad (5.7)$$

for cases (5.4a, b and c), respectively. Clearly, the range-versus-time curve for all three cases belongs to the type 3 pattern.

We now apply the above result to the case of cheatgrass introduced in section 2.4 (Fig. 2.11). By taking the square root of the area in Fig. 2.11(f) divided by $\sqrt{\pi}$, we obtain the effective range radius against time as shown in Fig. 5.4. We see that the plot is fitted by $5(t - 1890)$ in the beginning and it tends to increase exponentially in the later stages. Since our mathematical model predicts that the initial rate of range expansion is c, we have an estimate of $c = 5$ km/year as the rate of short-distance dispersal. As Mack (1981) noted, seeds of the cheatgrass were disseminated not only by animals but also by transportation facilities such as trains and horse carriages. If dispersal due to transportation facilities plays a major role in the long-distance dispersal, the seeds will disperse from almost anywhere within colonized areas. Then, case (c), $\lambda(r) = \lambda_2 r^2$, may be appropriate to describe the colonization rate. If we adopt $\lambda_2 = 3.5 \times 10^{-5}$/(km^2 year) for the colonization coefficient, the data points in Fig. 5.4 fit fairly well with a theoretical curve derived from eqn (5.6c), which is

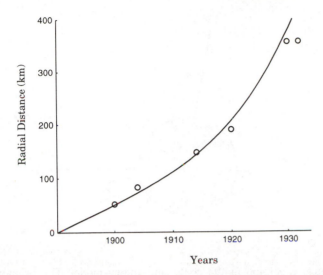

Fig. 5.4 Effective range radius versus time for cheatgrass. Solid line is the theoretical curve derived from eqn (5.6c) for $c = 5$ km/year and $\lambda_2 = 3.5 \times 10^{-5}$/(km^2 year); ○, data taken from Mack (1981) (see Fig. 2.11f).

shown by the solid curve. This means that cheatgrass generates only 3.5×10^{-5} successful nuclei per square kilometre in the invaded area per year. It is surprising that such a small amount of nucleation was sufficient to cover the entire western highlands of North America within two decades.

5.3 Coalescing colony model

In this section, we consider the case when departing migrants do not travel a great distance away from the parent population, and thus their colonies coalesce with the expanding parent population before long. The cases of red deer and the Himalayan thar in New Zealand, some avian species such as the European starling and house finch in North America and the rice water weevil can be considered to fall into this category. A general treatment of such situations would require a highly complex equation that incorporates the relative positions of the scattered colonies. By considering an idealized situation as below, however, this problem can be discussed within the basic framework of modelling developed in the previous section.

Assume that at $t = 0$ a few propagules invade a point and subsequently expand their range radius at a constant rate c by neighbourhood diffusion (see Fig. 5.5). We shall call this original group the primary population, and

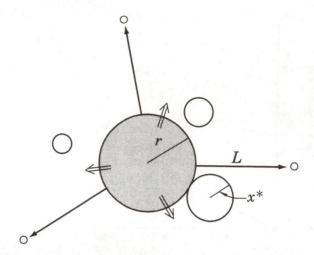

Fig. 5.5 Range expansion in the coalescing colony model. The primary colony (shaded circle) expands its range by short-distance dispersal and at the same time emits long-distance migrants at rate $\lambda(r)$, which settle at distance L ahead of the front of the primary population. Each offspring colony which expands at rate c coalesces with the primary population when their ranges overlap. Upon coalescence, the range of the primary colony including the offspring colony is immediately reshaped in a circular pattern, keeping the total area unchanged.

its range the primary colony. The primary population will sporadically generate long-distance migrants that will travel outward a constant distance L from the front of the primary colony to establish a new colony, which we shall call an offspring colony. The colonization rate is given by $\lambda(r)$ as defined in the previous section, where r is the range radius of the primary colony. Each offspring colony will immediately commence expanding concentrically at the constant rate c. An offspring colony will eventually collide with the primary colony and become part of the expanding primary colony. Here we assume that, upon every event of collision, the range of the combined colony is immediately reshaped in a circular pattern around the centre of the primary colony with the total area kept unchanged. We also assume that neither coalescence between offspring colonies nor secondary colonization by long-distance migrants originating in an offspring colony occurs. These assumptions may be permissible if the jump distance L is small enough that each offspring colony remains small until it coalesces.

What is the rate at which the area of the primary colony will expand? With this question in mind, we now proceed to formulate the process we have just described. If we denote by $r(t)$ the radius of the primary colony at time t and by $\rho(x,t)$ the density function of offspring colonies with radius x (using x in distinction to $r(t)$, the radius of the primary colony) at time t, we have an equation similar to eqn (5.2):

$$\frac{\partial \rho(x,t)}{\partial t} + \frac{\partial c \rho(x,t)}{\partial x} = 0 \qquad (x^*(t) > x > 0), \qquad (5.8)$$

where $x^*(t)$ is the radius of the offspring colony that is about to merge with the primary colony at time t, as shown in Fig. 5.5. Obviously, the sizes of offspring colonies, x, lie between 0 and $x^*(t)$. The initial and boundary conditions for eqn (5.8) are respectively given as follows:

$$\rho(x,0) = 0, \qquad (5.9)$$

$$c\rho(0,t) = \lambda(r). \qquad (5.10)$$

Condition (5.9) means that there are no offspring colonies at the outset, while condition (5.10) indicates that the production rate of offspring colonies with zero radius is equal to $\lambda(r)$, the colonization rate by migrants originating in the primary population. The solution of eqn (5.8) under the conditions (5.9) and (5.10) is given by

$$\rho(x,t) = \begin{cases} 0 & (x > ct > 0) \\ \dfrac{1}{c} \lambda(r(t-x/c)) & (ct \geq x) \end{cases}. \qquad (5.11)$$

Next we consider the coalescence process between the primary and offspring colonies. Since we have assumed that, as soon as the primary and offspring colonies come into contact, the former will incorporate the latter and return to a disc shape while maintaining the total area, we have the following equation:

$$\frac{d}{dt} \pi r^2 = \begin{cases} 2\pi rc & (t_s > t > 0) \\ 2\pi rc + \pi x^{*2} \rho(x^*,t)(c - \dot{x}^*) & (t > t_s) \end{cases}, \qquad (5.12)$$

where $t_s = L/2c$, and $\dot{x}^*$ denotes a first-order derivative with respect to time. The left-hand side represents the rate of change of the area of the primary colony. If the primary population emits long-distance dispersers immediately after it starts to expand at $t = 0$, both the primary and offspring colonies independently expand at a rate c until the first collision occurs at time $t_s = L/2c$. The upper equation of (5.12) describes area expansion solely by neighbourhood diffusion before the first collision occurs. When t exceeds t_s, the increase in area is caused not only by neighbourhood diffusion, but also by coalescence of offspring colonies as given by the additional term in the lower equation (see section 5.6).

Finally we determine the relation between the range radius $r(t)$ of the

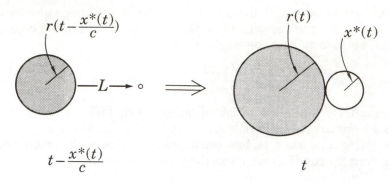

Fig. 5.6 Growth of primary and offspring colonies prior to their coalescence; $r(t)$ is the radius of the primary colony at time t, and $x^*(t)$ is the radius of the offspring colony that collides with the primary colony at time t. Since this offspring population has been expanding at the constant rate c, it was generated at time $t - x^*/c$ from the primary colony whose radius then was $r(t - x^*/c)$. (After Shigesada *et al.*, 1995.)

primary colony and the maximum radius $x^*(t)$ of offspring colonies at time t. Because the offspring colony has been expanding at a rate c, we know that its originating population left the primary colony at time $t - x^*/c$ (see Fig. 5.6). Since the radius of the primary colony at that time was $r(t - x^*/c)$, it has expanded by $r(t) - r(t - x^*/c)$, while the offspring colony has expanded by x^*. Since L is the sum of these two distances, we have the following equation:

$$L = r(t) - r(t - x^*/c) + x^*(t) \quad (t > t_s). \tag{5.13}$$

Note that $x^*(t)$ is smaller than $L/2$ and generally changes with time t.

We now have a set of equations (5.11)–(5.13) for range expansion with coalescence, which involves two variables, $r(t)$ and $x^*(t)$. If we differentiate eqn (5.13) with respect to t and substitute the resultant equation and (5.11) into (5.12), we have alternatively a set of differential equations with respect to $r(t)$ and $x^*(t)$ (see section 5.6):

$$\dot{r} = \frac{2r(c + \dot{r}') + x^{*2}\lambda'}{2r(c + \dot{r}') - x^{*2}\lambda'} c,$$

$$\dot{x}^* = \frac{c}{c + \dot{r}'} (\dot{r}' - \dot{r}), \tag{5.14}$$

where $\dot{r}' = \dot{r}\left(t - \dfrac{x^*(t)}{c}\right)$ and $\lambda' = \lambda\left(r\left(t - \dfrac{x^*(t)}{c}\right)\right)$.

As in the previous section, we examine eqns (5.14) for the following

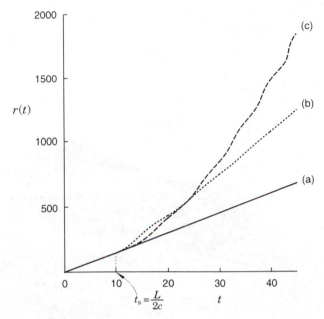

Fig. 5.7 Range-versus-time curve obtained from the coalescing colony model. Range radius of the primary colony, $r(t)$, is calculated for three types of colonization rate: (a) $\lambda(r) = 0.01$, $r(t)$ increases mostly at rate c, showing the expansion curve of type 1; (b) $\lambda(r) = 0.005r$, $r(t)$ shows biphasic expansion, type 2; (c) $\lambda(r) = 0.00003r^2$, $r(t)$ increases at an accelerating rate, type 3; $c = 15$ km/year, $L = 300$ km.

three cases with respect to the colonization rate from a primary colony of radius r:

$$\text{(a)} \quad \lambda(r) = \lambda_0, \qquad \text{(b)} \quad \lambda(r) = \lambda_1 r, \qquad \text{(c)} \quad \lambda(r) = \lambda_2 r^2.$$

We first numerically solve eqn (5.14). The results showing how the primary colony's radius $r(t)$ changes with time are given in Fig. 5.7. For the initial period $0 < t < t_s$, there is no coalescence and so $r(t)$ increases at the constant rate c for all three cases. After t_s, however, the curve of $r(t)$ exhibits different patterns depending on the form of $\lambda(r)$, and these features are summarized below.

In case (a), where long-distance migrants are generated at a constant rate regardless of the primary colony's size, the rate of spread is mostly constant at c, although it increases slightly after t_s. This is basically the type 1 pattern.

In case (b), where the long-distance migrants jump mainly from the primary colony's front, the spread rate shifts to a higher constant rate at

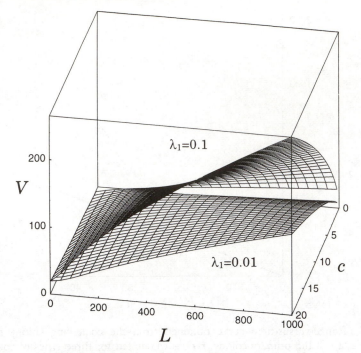

Fig. 5.8 Relation among ultimate speed, V, rate of spread by short-distance dispersal, c, and leap distance, L, for $\lambda_1 = 0.01$ and 0.1.

time t_s, thus exhibiting the type 2 pattern. The higher rate after t_s, $dr(\infty)/dt$, which we now denote by V, can be determined from eqns (5.14) as (see section 5.6):

$$2(V-c)(V+c)^2 - \lambda_1 c^2 L^2 = 0. \tag{5.15}$$

Figure 5.8 illustrates how the ultimate velocity V varies with the basic parameters c, L and λ_1 of the model.

In case (c), where the production rate of long-distance migrants is proportional to the primary colony's area, $r(t)$ exhibits the accelerating pattern of type 3, as shown in Fig. 5.7(c). Although the curve (c) shows a gentle undulation, this may result from the assumption that the long-distance migrants arrive only at a distance L from the primary colony.

Although we discussed only three basic forms of $\lambda(r)$ above, we can generalize as follows: range expansion occurs in an accelerating manner and so belongs to the type 3 pattern when $\lambda(r)$ is expressed as a function of a higher order of r than unity (i.e., $\lambda(r) \propto r^\theta$ with $\theta > 1$); if, on the other hand, $\lambda(r)$ is of a lower order of r than unity (r^θ with $\theta < 1$), range expansion occurs at a constant rate and hence belongs to the type 1 pattern. The type 2 pattern occurs only when $\lambda(r)$ is proportional to r, that is, when new colonies are founded mainly by long-distance migrants that are born at the front or perimeter of the primary colony (see section 5.6).

As stated earlier, for sake of analysis, several simplifications have been made in the present coalescing colony model. Thus, we assumed that the leap distance of a long-distance migrant is fixed at L, and also that the offspring colony is absorbed by the primary colony as soon as the two come into contact, with the primary colony always maintaining a circular range pattern. To see whether such assumptions are justifiable, a computer simulation was carried out for the case where long-distance migrants are stochastically generated at the rate $\lambda(r)$, and their leap distance is a random variable taken from a 'truncated' Gaussian distribution, that is, a probability density by excluding regions outside of the interval $[0, 2L]$ from a Gaussian distribution with mean L and variance σ^2. Furthermore, offspring colonies were assumed to continue spreading outward in a concentric manner even after coming into contact with the primary colony (see Fig. 5.9). We found that the resulting range-versus-time curves could

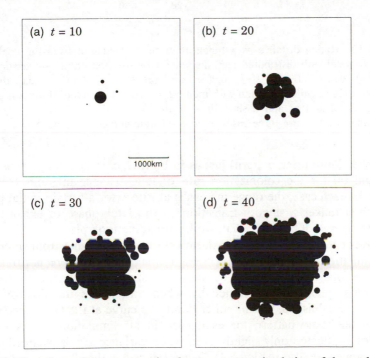

Fig. 5.9 Snapshots of range expansion from computer simulation of the coalescing colony model. Long-distance dispersal occurs stochastically at rate $\lambda(r) = 0.005r$, and its leap distance follows a truncated Gaussian distribution with mean $L = 300$ and standard deviation $\sigma = 40$, $c = 15$.

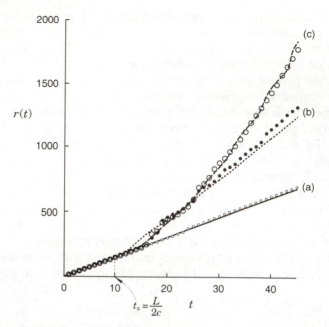

Fig. 5.10 Radial distance as a function of time calculated from the coalescing colony model with distributed leap distance. Colonization occurs stochastically at rate (a) $\lambda(r) = 0.01$, (b) $\lambda(r) = 0.005r$, and (c) $\lambda(r) = 0.00003r^2$. Leap distance follows a Gaussian distribution with mean $L = 300$ and standard deviation $\sigma = 40$, $c = 15$. Solid, dotted and dashed lines represent solutions of eqns (5.14) corresponding to the above three cases for $\sigma = 0$ (same curves as in Fig. 5.7).

be divided into three patterns just as in Fig. 5.7, corresponding to cases (a), (b) and (c) for the colonization rate of long-distance migrants (see Fig. 5.10). In each case, the range increases at rate c for a certain initial period and then makes a smooth transition to the later phase of expansion. It should be noted, however, that, while spreading took place at the constant rate c for the initial stage, this duration varied somewhat from simulation to simulation even when the same parameter values were used. This is because long-range dispersal occurs stochastically so that the duration of the initial phase is determined by when and where the first offspring colony is established. On the other hand, the curve at the later stage shows nearly the same pattern for each case in all simulations. Moreover its pattern is affected only slightly even when variance σ^2 is changed over a wide range (see also section 5.4). Overall, it seems that the assumptions we made in the analytical model provide a reasonably good approximation of the coalescing process.

We furthermore carried out a computer simulation for the case where

Table 5.1 Classification of range expansion patterns depending on colonization rate and leap distance

Colonization rate	Leap distance, L	
$\lambda(r) \propto r^\theta$	Moderate	Long
$\theta < 1$	type 1	type 3
$\theta \approx 1$	type 2	type 3
$\theta > 1$	type 3	type 3

offspring colonies also scatter long-distance migrants to create new colonies. While we again obtained three types of invasion curves similar to those in Fig. 5.7, corresponding to cases (a), (b) and (c) for the colonization rate, the spread rate obtained from computer simulation was clearly higher than that from the analytical model. For example, in case (b), the colonization coefficient λ_1 in the computer simulation should be lowered by one order if the spread rate from computer simulation is to be matched with the analytical one given by eqn (5.15), all other parameter values being the same.

As in the scattered colony model in the previous section, we neglected the establishment period immediately following invasion. The spread rate will clearly be slower than the ones obtained above if this effect is taken into consideration. Thus, various factors should be taken into account in quantitatively discussing the range-versus-time curve, but the stratified diffusion models introduced in the present and preceding sections should provide a basic framework for exploring the mechanisms underlying range expansion. Qualitative features obtained from the scattered colony model and coalescing colony model are summarized in Table 5.1.

5.4 Application: European starling and house finch

If actual data which show the expansion pattern of type 2 are available, we can apply the results presented in the previous section to estimate the three parameters—c, L and λ_1—that specify the basic invasion process. Thus, the spread rate c is equal to the slower initial constant rate in the curve. L can be determined from the time t_s when the shift in spread rate occurred, using $L = 2ct_s$. Meanwhile, if we estimate V by reading the higher spread rate after t_s, and substituting it together with c into eqn (5.15), we obtain the colonization coefficient λ_1.

Applying the above method to the spread curve of Fig. 2.5(b) for the European starling, as introduced in section 2.2, we obtain $c = 11$ km/year, $L = 300–400$ km ($t_s = 14–18$ year), $V = 50$ km/year and $\lambda_1 = 0.02/($km year). The role of long-distance migrants in this species is performed by the overwintering young birds, which are indicated by the black dots in Fig. 2.5(a). Although their actual ranges are obscure, they seem to lie in a band roughly 800 km wide surrounding the primary range. The mean length of the long-distance migration could be about half this band width, which reasonably coincides with the estimated value $L = 300–400$ km.

Next we apply our method to the range expansion of the house finch, which was extensively studied by Mundinger and Hope (1982) based on documented surveys of Christmas Bird Counts (see section 2.2). This species also exhibits the type 2 expansion pattern, with an initial slow spread (3.5 km/year) followed by a linear expansion at a higher rate (20.7 km/year). The distribution of the leap distance of this species is given by Fig. 2.7, in which the mean leap distance is 115 km and the standard deviation is 66 km. Thus we first substitute $L = 115$, $c = 3.5$ and $V = 20.7$ into formula (5.15) to obtain $\lambda_1 = 0.12/$(km year) for the colonization coefficient in the analytical model. Meanwhile, we carried out a computer simulation using the same parameter values as above (i.e., $c = 3.5$ and $\lambda_1 = 0.12$) and assuming the leap distance to be a random variable exhibiting truncated Gaussian distribution with mean $L = 115$ and various standard deviations in the range $\sigma = 10–100$. Figure 5.11 shows the range expansion with time calculated from the computer simulation for $\sigma = 60$

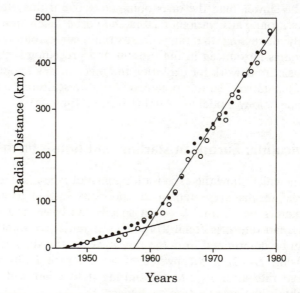

Fig. 5.11 Range expansion for house finch calculated from the coalescing colony model with distributed leap distance. Parameters chosen are $c = 3.5$ km/year, $\lambda(r) = 0.12r$, $L = 115$ km and $\sigma = 60$ km. Solid dots, calculated; circles, observed. Regression lines of calculated data indicate 3.5 km/year for the initial stage of spread and 21.5 km/year for the later stage.

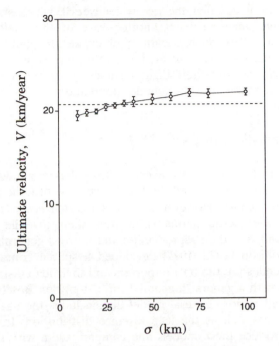

Fig. 5.12 Ultimate velocity of range expansion for house finch versus variance of leap distance. Leap distance follows a truncated Gaussian distribution with mean L and variance σ^2 confined in $[0, 2L]$. Each circle represents the average of the ultimate velocity obtained from 10 simulations, with the standard deviation indicated by the bars. Velocity remains relatively constant with changes in σ. Parameters are chosen as $c = 3.5$ km/year, $\lambda(r) = 0.12r$, $L = 115$ km. The dashed line indicates the ultimate velocity V calculated from the coalescing colony model, eqn (5.15).

(solid dots). The regression lines indicate 3.5 km/year for the initial stage of spread and 21.5 km/year for the later stage, which closely fit the observed data of 3.5 km/year and 20.7 km/year (circles), respectively. As noted in the previous section, the period of the early stage of spread varies somewhat between simulations. On the other hand, the spread rate in the later stage remains almost constant, even when the standard deviation σ is relatively large as shown in Fig. 5.12.

When the European starling and house finch are compared, the former has larger parameter values in both flight length L and spread rate c due to neighbourhood diffusion, whereas the colonization coefficient of the European starling is smaller by one order than that of the house finch. Overall, the ultimate spread rate of the European starling, $V = 50$ km/year, is 2.5 times that of the house finch.

Finally we point out that the rice water weevil, which shows a type 3 expansion curve, may represent another case of the coalescing colony model (see Fig. 2.10). Such a pattern of expansion could result if the long-distance migrants are generated at a rate proportional to the area of the primary colony and their flight distance is moderate, so that new colonies quickly merge into the founder population.

5.5 Mechanistic models

In the Skellam/Fisher models, every individual is assumed to move at random throughout its life, and the age structure within the population is not taken into account. However, as van den Bosch *et al.* (1990) pointed out, some species settle permanently after their juvenile period at a breeding ground, and their reproduction and survival generally depend on their age. Mollison (1972, 1977) developed a spatial contact model, in which the juveniles are the only dispersers and their leap distance assumes a distribution with a general functional form. Van den Bosch *et al.* (1990, 1992) presented a more general model that includes the age structure of the population as well as the leap distance distribution. In this section, we briefly introduce these models and compare them with the Skellam/ Fisher models or the stratified diffusion models presented in the preceding sections.

5.5.1 *Spatial contact models*

First we introduce the spatial contact model developed by Mollison (1972, 1977), which was originally used to explain the spread of epidemics.

Consider a population in one- or two-dimensional space, in which only juveniles undergo dispersal: an offspring is allowed only one move from the parent, or parents breed their offspring at a different location from their settling site. Let us denote the probability distribution for the dispersal distance of juveniles by $dV(s)$, which is referred to as the 'contact distribution' (this name originates in epidemiology); $dV(s)$ can be of any functional form including an exponentially bounded distribution that has long tails. If each parent produces offspring at a rate α, and the offspring instantaneously disperse from the parent's location, the recruitment rate of juveniles at position x is proportional to a weighted average over all possible parents; thus we have the following equation:

$$\frac{dn(x,t)}{dt} = \alpha \bar{n}, \tag{5.16}$$

where $n(x, t)$ is the population density at position x and time t, and $\bar{n}$ is a convolution integral taken over the space R as

$$\bar{n} = \int_R n(x - s, t) \, dV(s). \tag{5.17}$$

When the density effect is taken into account, eqn (5.16) is extended to a nonlinear integral equation:

$$\frac{dn}{dt} = \alpha \bar{n} \left(1 - \frac{n}{K} \right). \tag{5.18}$$

For the particular case of double exponential contact distribution along a line with standard deviation σ:

$$dV(s) = \frac{1}{\sqrt{2}\,\sigma} \exp(-\sqrt{2}\,|s|/\sigma) \, ds, \tag{5.19}$$

Mollison (1972) found that eqn (5.18) has travelling wave solutions of the form $n(x - ct)$ as in the Fisher equation, and the minimum wave velocity is given by

$$c_{min} = \frac{3\sqrt{3}}{2\sqrt{2}} \alpha\sigma = 1.873\alpha\sigma. \tag{5.20}$$

He also suggested that the solution starting from a localized distribution asymptotically approaches a travelling wave with the minimum velocity as above.

As for the linear equation (5.16), Mollison further found a wave solution in explicit form as $n(x, t) = A \exp\{-w(x - ct)\}$, and showed that its minimum velocity c_{min} is given by

$$c_{min} = \min_{w > 0} \frac{\psi(w)}{w}, \tag{5.21}$$

where

$$\psi(w) = \alpha \int_R \exp(ws) \, dV(s).$$

Using this formula, he determined the minimum velocity for various types of contact distributions. For example, when the contact distribution is symmetric with variance σ^2, the minimum wave velocity in one dimension is $\alpha\sigma\sqrt{e} = 1.649\alpha\sigma$ for a normal distribution, $1.568\alpha\sigma$ for a continuously uniform one, and $1.509\alpha\sigma$ for a distribution concentrated at $\pm\sigma$ (Daniels, 1975). Taking into account the fact that eqn (5.21) is unavailable if $\psi(w)$ diverges for all $w > 0$ and that the double exponential contact distribution given by eqn (5.19) is in a sense the borderline case, we can say that if the contact distribution $dV(s)$ has longer tails than an exponentially bounded distribution, the velocity of the wave propagation will be asymptotically infinite. Therefore, for given α and σ, the wave speed for a contact distribution with exponentially bounded tails as given by eqn (5.20) should be highest. Mollison also suggested that the minimum velocity in the non-linear contact model (5.18) is given by the same formula as (5.21) for the linear case (5.16), while the stochastic version of the non-linear contact model shows quite different features from the deterministic models above (Mollison and Daniels, 1993).

5.5.2 *Age-structured model*

When a population is age-structured, and the reproduction, survival and dispersal of individuals depend on their age, van den Bosch *et al.* (1990, 1992) presented a more general model to analyse range expansion from life-history parameters. They obtained approximate formulae for the velocity of expansion, which were applied to several biological invasions.

Consider a homogeneous environment in two-dimensional space, in which dispersal of the population has no preferred direction (i.e., it is rotationally symmetric). Let $b(t, x)$ denote the number of births per unit area and time at position x and time t. This quantity equals the sum of all current births at x from parents of all positive ages born at all possible places. Thus we have the following integral (renewal) equation:

$$b(t, x) = \int_R \int_0^\infty b(t - a, s) B(a, x - s)\, \mathrm{d}a\, \mathrm{d}s, \qquad (5.22)$$

where $B(a, x - s)$ is the expected rate of reproduction at position x by an individual of age a born at s. $B(a, x - s)$, termed the 'reproduction-and-dispersal kernel', incorporates the demographic and dispersal characteristics of the species, by putting

$$B(a, x - s) = L(a) m(a) D(a, x - s : \text{alive}), \qquad (5.23)$$

where $L(a)$ is the age-specific survivorship, defined as the probability that an individual is still alive at age a, and $m(a)$ is the age-specific fertility, defined as the rate of offspring production of an individual of age a. The dispersal characteristic is given by the conditional dispersal density, $D(a, x - s : \text{alive})$, which is the probability that an individual born at s is living at x at age a, given that it is still alive. If $b(t, x)$ is solved from eqn (5.22), the density of individuals living at x and time t is calculated from

$$n(t, x) = \int_R \int_0^\infty b(t - a, s) L(a) D(a, x - s : \text{alive})\, \mathrm{d}a\, \mathrm{d}s. \qquad (5.24)$$

Although eqn (5.22) with (5.23) can be applied to any particular species by choosing the appropriate functions describing its life-history characteristics, such an analysis would be complicated in general. Thus van den Bosch *et al.* (1990) investigated the velocity of population expansion by assuming that eqn (5.22) had travelling wave solutions of the form

$b(x - ct) = A \exp\{-k(x - ct)\}$, as in the case of the Fisher model (or the spatial contact model), to obtain approximation formulae for the velocity. They first analysed the special case that all individuals undergo random movements during their whole life, so that the dispersal density, $D(a, x - s : \text{alive})$, is given by a Gaussian density with variance σ^2 as

$$D(a, |x - s| : \text{alive}) = \frac{1}{2\pi\sigma^2 a} \exp\left(\frac{|x - s|^2}{2\sigma^2 a}\right), \qquad (5.25)$$

where $|x - s|$ indicates the Euclidean distance between x and s. In this situation, the velocity of population expansion is calculated as

$$c = \sqrt{2r\sigma^2}, \qquad (5.26)$$

where r, the intrinsic growth rate in the presence of age structure, is calculated from the Euler equation (Roughgarden, 1979):

$$1 = \int_0^\infty e^{-ra} L(a) m(a) \, da. \qquad (5.27)$$

If we put $r = \varepsilon$ and $\sigma^2 = 2D$, eqn (5.26) equals the velocity $c = 2\sqrt{\varepsilon D}$ calculated from the Fisher/Skellam model. Thus formula (5.26) is regarded as a variant of Fisher/Skellam's velocity in which age structure is taken into account. In fact, in section 3.7, we estimated the intrinsic growth rates ε for various species, some of which were calculated on the basis of eqn (5.27) or its discretized version.

For more general functions of the dispersal density, van den Bosch *et al.* derived the following approximation formula for the wave velocity:

$$c \approx \frac{\sigma}{\mu} \sqrt{2 \ln R_0} \left[1 + \left\{\left(\frac{\nu}{\mu}\right)^2 - \beta + \frac{1}{12} \gamma\right\} \ln R_0\right], \qquad (5.28)$$

where

$$R_0 = \int_0^\infty L(a) m(a) \, da,$$

$$\mu = \frac{1}{R_0} \int_0^\infty a L(a) m(a) \, da,$$

$$\sigma^2 = \int_{-\infty}^\infty \int_0^\infty \int_{-\infty}^\infty x^2 L(a) m(a) D(a, x, s : \text{alive}) \, ds \, da \, dx,$$

$$\nu^2 = \frac{1}{R_0} \int_0^\infty a^2 L(a) m(a) \, da - \mu^2,$$

$$\gamma = \frac{1}{\sigma^4} \int_{-\infty}^\infty \int_0^\infty \int_{-\infty}^\infty x^4 L(a) m(a) D(a, x, s : \text{alive}) \, ds \, da \, dx - 3,$$

$$\beta = \frac{1}{\gamma^2 \mu R_0} \int_{-\infty}^\infty \int_0^\infty \int_{-\infty}^\infty x^2 a L(a) m(a) D(a, x, s : \text{alive}) \, ds \, da \, dx - 1.$$

R_0 represents the net reproduction; μ and ν^2 are respectively the mean and variance of age at child-bearing; σ^2 and γ are the variance and kurtosis of the marginal dispersal density; and β is a measure of the interaction between dispersal and reproduction. This formula was shown to be valid for $R_0 \le 7$ and $\nu/\mu \le 0.6$. Furthermore, when $R_0 < 1.5$, the second term within the brackets in eqn (5.28) becomes negligible, so that we have

$$c \approx \frac{\sigma}{\mu} \sqrt{2 \ln R_0} \quad \text{for} \quad R_0 < 1.5. \qquad (5.29)$$

It should be noted that eqn (5.29) has the same form as the velocity obtained from the Fisher/Skellam model if we use the following identifications:

$$\ln R_0 = \varepsilon\mu \quad \text{and} \quad \sigma^2 = 2D\mu. \qquad (5.30)$$

These relations may roughly be justified if we take the mean age at child-bearing as the unit of time. Thus formula (5.29) is another variant of the Fisher/Skellam velocity.

Using these formulae, van den Bosch *et al.* (1990, 1992) and Metz and van den Bosch (1995) analysed the spread of birds (collared dove, starling in Europe and USA, cattle egret, house sparrow), mammal species (muskrat for 1900–30 and 1930–60) and epidemics of fungi (downy mildew foci on spinach, stripe rust foci on wheat) and rabies. For some species, they collected field data on the demographic parameters and information from mark–recapture experiments to construct the dispersal density. Based on those data, they calculated the velocities of range expansion from both eqns (5.28) and (5.29). The calculated values fitted well with the observed velocity (see Fig. 5.13). They also remarked that formula (5.28) gives in all cases except one a slightly better prediction for the velocity than does formula (5.29), though the improvement is not statistically significant ($P \approx 0.1$). Thus we may conclude that the variant of Fisher/Skellam velocity given by eqn (5.29) can be used satisfactorily in many cases.

Finally we compare the stratified diffusion models (scattered colony model and coalescing colony model) presented in sections 5.2–5.3 with the mechanistic models (spatial contact model and age-structured model) described in the present section. In the stratified diffusion models, we focus on the patch size distribution, whereas both the contact model and age-structured model deal with the spatial distribution of the population density. More precisely, in the stratified diffusion models, we ignore the population dynamics within each patch by assuming that the population density has already reached the carrying capacity everywhere in the patch. A similar idea was employed in the metapopulation model which we discussed in section 4.6. Thus, detailed information on the population

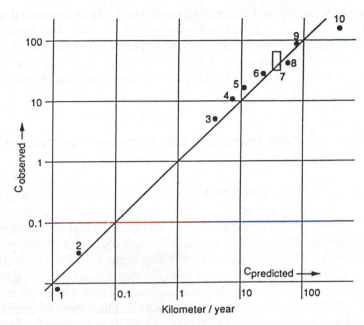

Fig. 5.13 Observed and predicted speeds of range expansion: 1, downy mildew foci on spinach; 2, stripe rust foci on wheat; 3, muskrat (Europe after 1930); 4, muskrat (Europe before 1930); 5, house sparrow (North America); 6, house sparrow (Europe); 7, rabies (Europe); 8, collared dove (Europe); 9, starling (North America); 10, cattle egret (South America). (After Metz and van den Bosch, 1995.)

density is sacrificed for the sake of simplicity in the stratified diffusion models. It should be noted, however, that a non-linear density effect is still incorporated particularly in the coalescing colony model, where we assumed that after two colonies coalesce, the population density within the overlapping area quickly reaches the carrying capacity.

Although we have treated short-distance and long-distance dispersals as distinct processes in the stratified diffusion models, they are incorporated at the same time into the dispersal kernel in mechanistic models. Therefore, if we have detailed data on the demographic and dispersal characteristics of a species, we should first apply the mechanistic models to obtain the asymptotic speed of a travelling wave. When the observed data on the rate of spread increases with time as in type 3, however, we have to seek for a solution that forms an acceleratingly progressing wave. Unfortunately, no mechanistic model for describing such a process seems to have been developed yet. Meanwhile, in the stratified diffusion models, the range-versus-time relation of type 1, 2, or 3 can be derived, depending on how and where short- and long-distance dispersers are generated. As mentioned before, we made many plausible simplifications in constructing the stratified diffusion model, so that the results are generally applicable only for qualitative purposes. None the less we think that a metapopulation-oriented model would be one promising approach if it could properly incorporate data on demographic characteristics in the model.

5.6 Appendix: the coalescing colony model—derivation of eqns (5.12)–(5.15)

Consider an offspring colony that just comes in contact with the primary colony at time t (see Fig. 5.5). Then offspring colonies expanding in isolation have radii smaller than $x^*(t)$. Among them, colonies of radius ranging between $x^*(t+dt) - cdt$ and $x^*(t)$ will coalesce with the primary one during $(t, t+dt)$, because the offspring upon collision at $t+dt$ has radius $x^*(t+dt)$, which has been growing at rate c from t to $t+dt$. Thus the total number of offspring colonies coalesced during $(t, t+dt)$ is given by $\rho(x^*, t)\{x^*(t) - (x^*(t+dt) - cdt)\} + O(dt^2)$. Therefore the increase of the primary colony's area by coalescence during dt is $\pi x^{*2}\rho(x^*, t)\{x^*(t) - (x^*(t+dt) - cdt)\} + O(dt^2)$. Dividing the increment by dt and tending dt to zero, we have $\pi x^{*2}\rho(x^*, t)(-dx^*(t)/dt + c)$ as given in eqn (5.12).

Next, we substitute eqn (5.11) into the right-hand side of eqn (5.12), and we obtain

$$\frac{d}{dt}r = \begin{cases} c & (t_s > t > 0) \\ c + \dfrac{x^{*2}}{2rc}\lambda(r(t - x^*/c))(c - \dot{x}^*) & (t > t_s) \end{cases} \tag{5.31}$$

Equations (5.31) and (5.13) together form a closed set of equations with respect to $r(t)$ and $x^*(t)$. Thus, differentiating eqn (5.13) with respect to time and combining it with eqn (5.31) will yield eqn (5.14).

Now, we determine the asymptotic solution for eqn (5.31) when $\lambda(r) = \lambda_1 r$. If we assume that dr/dt converges to V (constant) at $t \to \infty$, the second equation of (5.14) will yield $\dot{x}^* = 0$. Since $0 \le x^*(t) \le L/2$, $x^*(t)$ will converge to $x^*(\infty)$ (constant) when $t \to \infty$. Thus, for $t \to \infty$, eqns (5.31) and (5.13) can respectively be expressed as follows:

$$V = c + \frac{\lambda_1}{2}x^*(\infty)^2, \tag{5.32}$$

$$L = \frac{V}{c}x^*(\infty) + x^*(\infty). \tag{5.33}$$

By eliminating $x^*(\infty)$ from eqns (5.32) and (5.33), we obtain eqn (5.15).

For more general cases of $\lambda(r) = \lambda_\theta r^\theta (\theta > 0)$, we can make a rough estimation of the expansion rate. If $\dot{x}^* = 0$ for sufficiently large t, the second equation of (5.31) is approximated as

$$\frac{\mathrm{d}}{\mathrm{d}t} r \approx c + \frac{\lambda_\theta x^{*2}}{2} r^{\theta - 1}.$$

Thus if $\theta < 1$, $\mathrm{d}r/\mathrm{d}t$ tends to c as the range increases, so that the rate of spread approaches a constant rate c. If $\theta > 1$, on the other hand, $\mathrm{d}r/\mathrm{d}t$ increases with the range radius and hence the rate of spread increases acceleratingly with time.

6

Invasion of competing species

6.1 Competition between resident and invading species

So far we have focused our attention on the spread of an invading species without referring to its effects on the pre-existing ecosystem. In reality, however, an invading species will always interact with the resident species, often in the form of predation (including host–parasite interaction) or competition, at times even resulting in the extinction of the resident species.

In this chapter, we look at the case when an invading species is in competition with a resident species for habitat or food, and examine how the range boundary between the invading and resident species will shift as the invasion proceeds.

There are three possible strategies whereby an invading species can persist: (1) it takes over the resident species' habitat through direct competition; (2) competition with the resident species is not so aggressive that it coexists with the resident species after establishment; (3) even though competitively weak, it survives by moving into open spaces that arise either occasionally or periodically.

The first case occurs when the invading species spreads its range by displacing the resident species; successful alien species are often of this type, and sometimes the existing ecosystem is drastically altered because of this (the Argentine fire ant in the United States and many other examples are presented in Elton, 1958). In the third case, the species which is always moving from one open space to another is called a fugitive species; many annual plants and highly mobile bird species belong to this category. Invasion of fugitive species will be dealt with in Chapter 7.

Below, we analyse the process whereby an invading species spreads into an area occupied by a competing resident species by using the following set of equations, which is an extended form of the Fisher equation of (3.1):

$$\frac{\partial n_1}{\partial t} = D_1 \frac{\partial^2 n_1}{\partial x^2} + (\varepsilon_1 - \mu_{11}n_1 - \mu_{12}n_2)n_1,$$

$$\frac{\partial n_2}{\partial t} = D_2 \frac{\partial^2 n_2}{\partial x^2} + (\varepsilon_2 - \mu_{21}n_1 - \mu_{22}n_2)n_2, \tag{6.1}$$

where $n_1(x,t)$ and $n_2(x,t)$ are respectively the population densities of species 1 and 2, which represent the resident and invading species, respectively. The change in distribution range is caused by the diffusion term (first term on the right-hand side) and the growth term (second term). D_1 and D_2 are the diffusion coefficients, ε_1 and ε_2 are the intrinsic growth rates, and μ_{ii} and μ_{ij} $(i,j=1,2)$ are the coefficients of intra- and inter-specific competition; $\mu_{ij}n_j$ expresses the negative effect that the competition between species i and j has on the growth rate of species i.

6.2 The competition equation

Before analysing eqn (6.1), we first examine the case when two species compete while remaining in the same area without diffusive movement, that is, when competition takes place according to the so-called Lotka–Volterra competition equation, which is eqn (6.1) without the diffusion term:

$$\frac{dn_1}{dt} = (\varepsilon_1 - \mu_{11}n_1 - \mu_{12}n_2)n_1,$$

$$\frac{dn_2}{dt} = (\varepsilon_2 - \mu_{21}n_1 - \mu_{22}n_2)n_2. \tag{6.2}$$

Here we use a phase plane diagram to examine the trajectory of the solution in a qualitative manner. As shown in Fig. 6.1, we first draw null clines in the (n_1, n_2) space, which are obtained by setting the right-hand sides of eqns (6.2) to zero. So the null clines for species 1 are $n_1 = 0$ or $n_2 = \varepsilon_1/\mu_{12} - \mu_{11}n_1/\mu_{12}$ (the solid lines), and for species 2 $n_2 = 0$ or $n_2 = \varepsilon_2/\mu_{22} - \mu_{21}n_1/\mu_{22}$ (the dotted lines). Figure 6.1 shows the four possibilities for the relative positions of the null clines. The intersections of the null clines give the following four equilibrium points:

$E_0: (0,0)$, $E_1: (\varepsilon_1/\mu_{11}, 0)$, $E_2: (0, \varepsilon_2/\mu_{22})$, $E_3:$

$$\left(\frac{\varepsilon_1 \mu_{22} - \varepsilon_2 \mu_{12}}{\mu_{11} \mu_{22} - \mu_{12} \mu_{21}}, \frac{\varepsilon_2 \mu_{11} - \varepsilon_1 \mu_{21}}{\mu_{11} \mu_{22} - \mu_{12} \mu_{21}} \right). \tag{6.3}$$

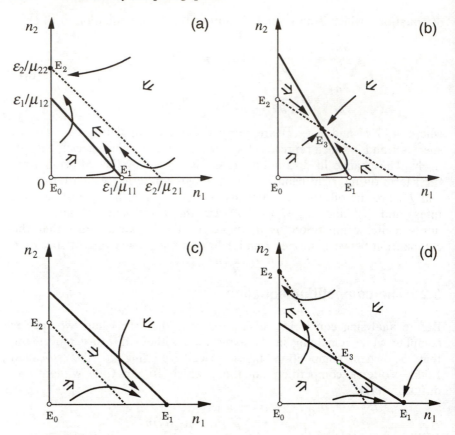

Fig. 6.1 Phase plane diagrams of Lotka–Volterra competition model. Solid lines, null clines of species 1; dotted lines, null clines of species 2; solid arrows, typical trajectories. Four possible cases are: (a) species 2 always wins; (b) both species coexist; (c) species 1 always wins; (d) either species 1 or species 2 wins depending on initial conditions. Circles, unstable equilibria; solid dots, stable equilibria.

The null clines divide the (n_1, n_2) plane into regions where dn_1/dt or dn_2/dt is either positive or negative; within each region the direction in which n_1 and n_2 change with time is indicated by a short arrow. Thus, the phase plane trajectory can be roughly depicted by following the arrow's direction. Typical trajectories are drawn by the solid arrows in Fig. 6.1: any trajectory approaches equilibrium point E_2 for (a), E_3 for (b), and E_1 for case (c), while in case (d) trajectories reach either E_1 or E_2, depending on the initial point. These results are reinterpreted in terms of competition:

(i) when $\dfrac{\varepsilon_1}{\mu_{11}} < \dfrac{\varepsilon_2}{\mu_{21}}, \dfrac{\varepsilon_2}{\mu_{22}} > \dfrac{\varepsilon_1}{\mu_{12}}$, only species 2 wins;

(ii) when $\dfrac{\varepsilon_1}{\mu_{11}} < \dfrac{\varepsilon_2}{\mu_{21}}, \dfrac{\varepsilon_2}{\mu_{22}} < \dfrac{\varepsilon_1}{\mu_{12}}$, both species coexist;

(iii) when $\dfrac{\varepsilon_1}{\mu_{11}} > \dfrac{\varepsilon_2}{\mu_{21}}, \dfrac{\varepsilon_2}{\mu_{22}} < \dfrac{\varepsilon_1}{\mu_{12}}$, only species 1 wins; and

(iv) when $\dfrac{\varepsilon_1}{\mu_{11}} > \dfrac{\varepsilon_2}{\mu_{21}}, \dfrac{\varepsilon_2}{\mu_{22}} > \dfrac{\varepsilon_1}{\mu_{12}}$, either species 1 or species 2 wins

depending on the initial conditions. (6.4)

Since we are interested in whether the few invading individuals of species 2 can become established in an area occupied by species 1, we need to determine whether n_2 will increase or not, starting from the vicinity of equilibrium point $E_1 : (\varepsilon_1/\mu_{11}, 0)$ at which only species 1 survives. A quick inspection of Fig. 6.1 reveals that n_2 increases only for cases (i) and (ii). In case (i), the invading species outcompetes the resident species, completely displacing it, while in case (ii), the solution approaches a stable equilibrium point given by E_3, where the competition remains balanced and the two species coexist.

6.3 Retreat of resident species

Returning to eqn (6.1), we now examine the invasion process when the invading and resident species are both able to undergo dispersal.

First, we assume that the entire area is occupied by the resident species (species 1) so that its population density has reached the carrying capacity $n_1(x, 0) = \varepsilon_1/\mu_{11}$ at all points. Into this situation, a few individuals of species 2 invade the vicinity of the origin. As discussed in section 4.3, for the invasion to be successful the population must increase when few in number. Mathematically this means that the equilibrium solution $(n_1(x), n_2(x)) = (\varepsilon_1/\mu_{11}, 0)$ of eqn (6.1) is unstable to perturbations. With some simple analysis, it can be shown that the conditions for instability are the same as in the case where no diffusion takes place, namely (i) or (ii) of (6.4).

We numerically solved eqn (6.1) by setting the initial condition as mentioned above and found that for cases (i) and (ii) of (6.4) the invasion proceeds as shown in Fig. 6.2(a) and (b), respectively. In both cases, the spread of the invading species (species 2) can be described by a travelling wave of constant speed whose front maintains a constant pattern, just as in the Fisher model. The rear of the travelling wave, however, differs for the two cases: in (a), the invading species completely displaces the resident one, whereas in (b), while the resident species' density decreases, competition between the two species eventually settles at the equilibrium state E_3 as given by (6.3), allowing them to coexist. Volpert *et al.* (1994) have

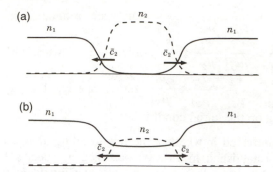

Fig. 6.2 Spread of invading species (species 2) in an area pre-occupied by species 1. (a) When species 2 always wins in the absence of diffusion, species 2 succeeds in invasion and establishes a travelling wave which proceeds by completely displacing the resident species. (b) When both species coexist in the absence of diffusion, species 2 can invade and establish a travelling wave, at the rear of which both invading and resident species coexist. In both cases, $\bar{c}_2$ is the speed of the travelling wave as given by eqn (6.6).

proved mathematically that, indeed, travelling waves always exist for cases (i) and (ii) of (6.4) (see also Hosono, 1989).

The speed of the travelling wave is heuristically derived in the following way. At the wave front of the invading species, the population densities of the two species are approximately given by $n_1 \approx \varepsilon_1/\mu_{11}$ and $n_2 \approx 0$, and so the second equation of (6.1) can be approximately expressed as

$$\frac{\partial n_2}{\partial t} = D_2 \frac{\partial^2 n_2}{\partial x^2} + \left(\varepsilon_2 - \mu_{21} \frac{\varepsilon_1}{\mu_{11}} \right) n_2. \tag{6.5}$$

This has the same form as the Skellam equation introduced in section 3.4. Thus, by letting $D_2 \to D$ and $\varepsilon_2 - \varepsilon_1 \mu_{21}/\mu_{11} \to \varepsilon$, eqn (6.5) becomes the one-dimensional version of eqn (3.16). Since the solution of the Skellam equation evolves into a travelling wave whose propagating speed is given by eqn (3.24) (for both the one- and two-dimensional versions), the same formula applied to eqn (6.5) yields

$$\bar{c}_2 = 2 \sqrt{\varepsilon_2 D_2 (1 - \varepsilon_1 \mu_{21}/\varepsilon_2 \mu_{11})}, \tag{6.6}$$

which is the speed of range expansion of species 2 into the resident species' habitat. To derive the speed more rigorously, we seek travelling wave solutions to eqn (6.1) of the forms $n_1(x - ct)$ and $n_2(x - ct)$ as we did for the Fisher equation (see section 3.9). From the requirement that $n_2(x - ct)$ is positive at the front of the wave, we can show that $\bar{c}_2$ as defined by eqn (6.6) should be the minimum speed of the travelling wave, and computer simulation suggests that this minimum speed is in fact the actual speed for a wide range of parameter values. It should be noted, however, that Hosono (1989) showed that formula (6.6) becomes inapplicable when D_2 or $1/\varepsilon_1$ is extremely small compared with other parameter values.

We next see how the presence of a competing resident species will affect the travelling speed of the invading species. If species 2 invades an area that is absent of any competing species, the distribution of species 2 will change according to eqn (6.1) with $n_1 = 0$:

$$\frac{\partial n_2}{\partial t} = D_2 \frac{\partial^2 n_2}{\partial x^2} + (\varepsilon_2 - \mu_{22} n_2) n_2.$$

Since this equation has the same form as the Fisher equation (3.1), range expansion of species 2 will occur in the form of a travelling wave with a speed of

$$c_2 = 2\sqrt{\varepsilon_2 D_2}. \tag{6.7}$$

Dividing $\bar{c}_2$ by c_2, we obtain

$$\bar{c}_2/c_2 = \sqrt{1 - \varepsilon_1 \mu_{21}/\varepsilon_2 \mu_{11}} \equiv \gamma < 1,$$

indicating that the presence of a resident competing species will lower the propagation speed by the factor γ. In particular, if the invading and resident species have similar growth rates and competitive capacities (i.e., $\varepsilon_1 \approx \varepsilon_2$ and $\mu_{21} \approx \mu_{11}$), γ will be close to zero, so that $\bar{c}_2$ will be much smaller than c_2. In other words, if the invading species has only a slight competitive advantage over the resident species, the displacement of the latter by the former will be an extremely slow process. For example, Kohyama and Shigesada (1996) used a size-structure-based model combined with seed diffusion to examine how latitudinal forest zonations would change in response to global environmental change. They found that a global warming of around 3°C over a 100-year period will cause the biomass in resident forests to change simultaneously in response to the warming, with the boundaries of different forest types moving at rates slower by one or two orders of magnitude as compared with the potential rates of expansion in the absence of competition.

6.4 Competition between grey and red squirrels

As an example of an invading species that completely displaced a resident species, we introduce the work by Okubo *et al.* (1989) on the spread of the grey squirrel in Britain after it was introduced from North America.

The North American grey squirrel (*Sciurus carolinensis*) was released from various sites in Britain around the turn of this century. Although Britain had, as an indigenous species, the red squirrel (*Sciurus vulgaris*), it began disappearing from areas which the introduced grey squirrel had invaded and where it had successfully spread its range (see Fig. 6.3)

(a)

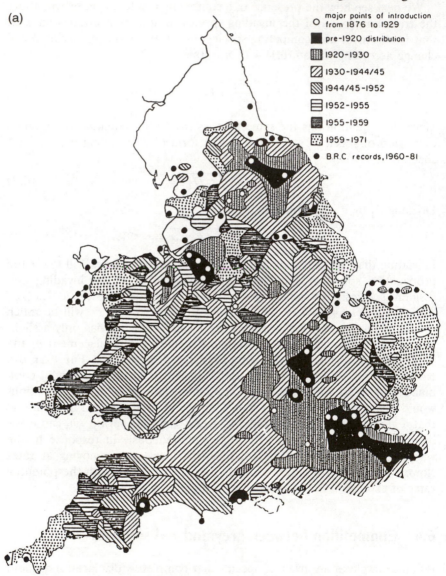

○ major points of introduction
  from 1876 to 1929

■ pre-1920 distribution

▥ 1920–1930

▨ 1930–1944/45

▧ 1944/45–1952

▤ 1952–1955

▦ 1955–1959

▦ 1959–1971

● B.R.C. records, 1960–81

(MacKinnon, 1978; Lloyd, 1983; Reynolds, 1985; Williamson and Brown, 1986). The native red squirrel is at a clear competitive disadvantage, with only half the body weight of the grey squirrel and a lower growth rate. In North America both species are present, though they occupy separate niches, with the red squirrel living mainly in the northern conifer forests and the grey squirrel mostly in mixed hardwood forests. In Britain, however, the absence of the grey squirrel had allowed the indigenous red squirrel to adapt to mixed hardwood forests as well to occupy an extensive

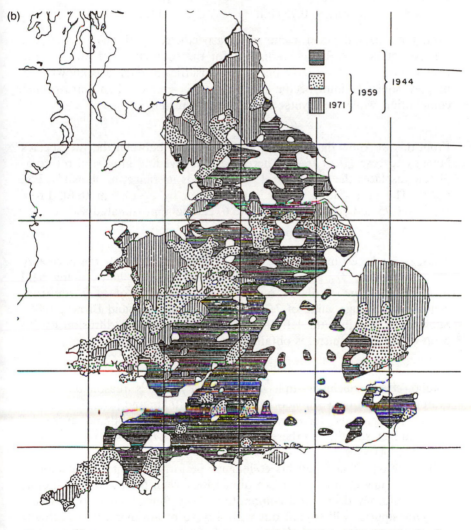

Fig. 6.3 (a) Spread of grey squirrel in England from 1920 to 1971. (b) Distribution decline of the red squirrel from 1944 to 1971. (After Lloyd, 1983.)

range. It seems that the red squirrel's range, particularly in the hardwood forests, was forced back when faced with the invasion by the grey squirrel, which was adapted to that particular environment.

With the red squirrel as species 1 and the grey as species 2, Okubo *et al.* used eqn (6.1) to analyse the grey squirrel's invasion process. Based on extensive field observations, the intrinsic growth rates and carrying capacities of the two species were estimated as follows:

$$\varepsilon_1 = 0.61/\text{year}, \qquad \varepsilon_2 = 0.82/\text{year},$$
$$\varepsilon_1/\mu_{11} = 0.75/\text{ha}, \qquad \varepsilon_2/\mu_{22} = 10/\text{ha}.$$

From these data, the coefficients of intraspecific competition, μ_{11} and μ_{22}, can be obtained. Although estimates for the coefficients of interspecific competition are not available because of insufficient data, it is known that in grey squirrels interspecific competition is greater than intraspecific competition, while the reverse holds in red squirrels, or

$$\mu_{12} > \mu_{22}, \, \mu_{11} > \mu_{21}.$$

From this we know that the competitive relation between the two species belongs to case (i) of (6.4). Therefore, the invading grey squirrel completely displaces the red squirrel as it spreads its range, as shown in Fig. 6.2(a). The propagation speed is then given by $\bar{c}_2$ of eqn (6.6). From $\varepsilon_1/\varepsilon_2 = 0.75$ and $\mu_{21}/\mu_{11} < 1$, eqn (6.6) can be approximated as

$$\bar{c}_2 \approx 2\sqrt{\varepsilon_2 D_2} = 2\sqrt{0.82 D_2}. \tag{6.8}$$

Thus, $\bar{c}_2$ can be estimated if D_2 is known, but unfortunately the necessary data to evaluate D_2 are not available. On the other hand, many field measurements are available on the invading speed of grey squirrels, yielding an average value of 7.7 km/year (Williamson and Brown, 1986). Substituting this into the left-hand side of eqn (6.8), the diffusion coefficient of the grey squirrel is obtained as

$$D_2 = 18 \text{ km}^2/\text{year}.$$

Okubo *et al.* concluded that $D_2 = 18$ km^2/year is reasonable if the average distance between woodlands is about 10 km and juvenile squirrels leave their parents' territory for a neighbouring woodland every 1.4 years. In fact, substituting these values into eqn (3.7) gives $D = \langle r^2 \rangle/4t = 10^2/(4 \times 1.4) \approx 18$.

In addition, Okubo and his colleagues performed numerical computation of the model in one and two dimensions when the expanding species meets uniformly distributed competitors. They found that in both cases, the grey squirrel will spread out with a wave of advance that eventually attains the constant speed given by eqn (6.6). As seen in the map given by Lloyd (Fig. 6.3), the grey was introduced at several sites. At each site, spread occurred, and some of the spreading waves eventually coalesced.

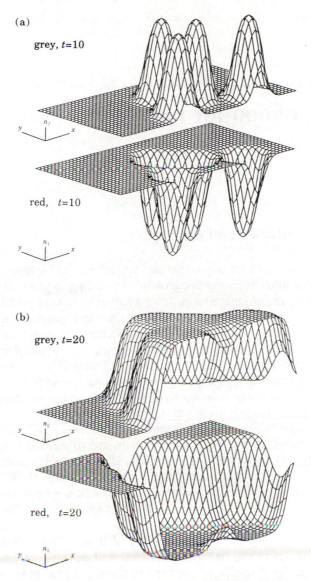

Fig. 6.4 Spread of grey squirrel in mathematical model (after Okubo *et al.*, 1989). The two-dimensional version of eqns (6.1) is numerically solved. Initial distribution consists of reds at the carrying capacity seeded with small pockets of greys at four points. (a) Surface plot of solution at $t = 10$: base density of greys is 0.0 and base density of reds is their carrying capacity. (b) At $t = 20$: greys spread outwards while reds recede, and eventually greys drive out reds.

Figure 6.4 presented by Okubo *et al.* (1989) visually depicts this process and explains the observed patterns of spread seen in Lloyd's map.

7

Competition for open space

7.1 Disturbance and open space

An open space, to an organism, can be defined as 'a space which when invaded will allow reproductive growth'. According to this definition, a new continent to an invading species from another continent could constitute a large open space. Likewise, freshly deglaciated grounds that appeared after the retreat of ice-sheets in the northern hemisphere during the Quaternary Period were also large open spaces. By analyzing the fossil pollen record taken from lake sediments, Davis (1981) showed that the open space (tundra) that appeared in North America after the last ice age (16,000 years ago) was consecutively invaded from the southeast by spruce (*Picea*), oak (*Quercus*), white pine (*Pinus*), hemlock (*Tsuga*), beech (*Fagus*) and somewhat later chestnut (*Castanea dentata*), etc., and that each took a different route, advancing its range northward or westward at the rate of about 100–400 m/year (Fig. 7.1; see also Bennett, 1983, 1986; Jacobson *et al.*, 1987). One of the most rapid migrants among them is oak (350 m/year), although the acorns from a parent oak are disseminated usually by only a few metres per year. In his perceptive article, Skellam (1951) suggested based on his model (see section 3.4) that small animals must have played a major role as the agents of long-distance dispersal of oak seeds. White pine and hemlock appeared first at refuge areas on the east coast or in the foothills of the Appalachians. White pine, which is less shade-tolerant than hemlock and grows more abundantly on disturbed sites or on poor soils, expanded rapidly northward and westward. On the other hand, hemlock, a slow-growing, shade-tolerant species, migrated at a speed of 200–300 m/year, arriving at most sites 500–1000 years later than white pine. It penetrated closed forests of pine, oak, birch, and maple, reaching maximum abundance in the northeast 5000 years ago, and declined in frequency in more recent years (Davis, 1976). Davis also pointed out that the glacial cycles were much shorter than had been previously thought, so that many species that were unable to adapt fast enough became extinct, and the ones that survived were mainly those species with high mobility or those that inhabited refuges which had escaped becoming frozen over. Changes in the location and abundance of beech populations (pollen percentages) were analysed by Dexter *et al.* (1987) by using a generalized Fisher model that incorporated an advection term (see section 3.6). They estimated the coefficients for diffusion and advection to be 9.6 km^2/year and 0.05 km/year, respectively.

There are also a variety of relatively small open spaces created by disturbances: areas that recently underwent a forest fire or flood, empty space in algal or sessile animal communities on rocky shores or coral reefs formed as a result of severe wave action during hurricanes or tidal waves (Sousa, 1979), gaps opened in the ground by a fallen tree (Denslow, 1987), or territorial vacuums which became available by the death or migration of the previous resident organism. The cycles in which such small open spaces are created and then disappear occur on a relatively short time scale (see Pickett and White (1985) for various other examples and DeAngelis *et al.* (1985) for a classification of mathematical models of disturbances).

The environment of an organism's habitat is thus often viewed as being in a state of non-equilibrium, stochastically changing in time and space. In fact, the 'non-equilibrium theory', which states that an unpredictably changing environment favours coexistence among many species, has attracted much attention in recent years (Sale, 1977; Huston, 1979; Chesson and Warner, 1981). In particular, Connell (1978) has noted the impressive biodiversity supported by coral reef ecosystems in the tropical Indo-Western Pacific, and proposed the 'intermediate disturbance hypothesis', which asserts that the highest diversity is maintained by intermediate levels of disturbance. At low levels of disturbance, better competitors get the chance to dominate the space, while at high levels of disturbance, many species will simply become extinct. Meanwhile, intermediate disturbances such as large waves or typhoons could make open spaces available to the competitively weaker species before the competitively stronger species is given enough time to dominate the entire ecosystem.

In the following sections, we examine the process by which the competition for space occurs when two competing species invade an open space one after the other; first for the case when an infinite expanse of open space is available, then the case when open spaces of relatively small sizes are sporadically created by disturbances.

7.2 Competition for large open spaces

As we saw in Chapter 6, when a species invades an area which is entirely

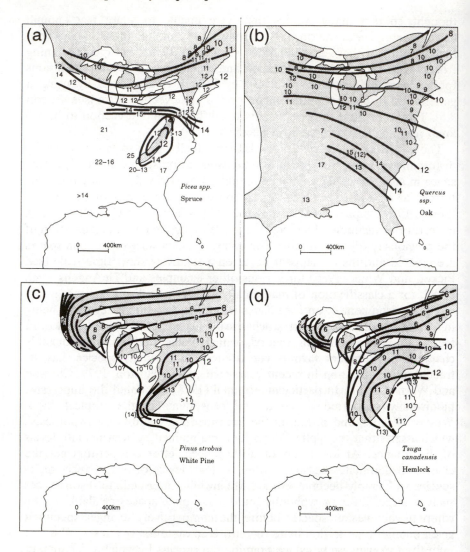

occupied by a resident species, the former must be competitively stronger than the latter in order to successfully establish itself. However, if two species consecutively invade an open area and the earlier species is still in the process of expanding its range, it may be possible for a competitively weaker species arriving later to successfully spread its range by reaching residual open spaces before the earlier invading species occupies the whole area. How large must the diffusive capability of the competitively weaker species be in order for it to successfully reach the open space? We

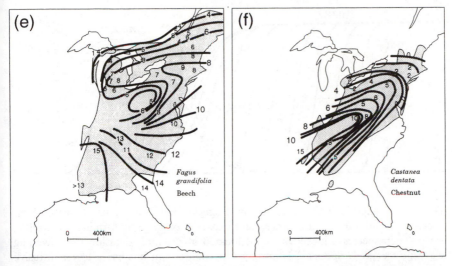

Fig. 7.1 Geographic spread of several tree species through North America after the last ice age (after Davis, 1981): (a) spruce; (b) oak; (c) white pine; (d) hemlock; (e) beech; (f) chestnut. The numbers refer to the radiocarbon age (in thousands of years) of the first appearance of the respective species at the site after 15,000 years BP. Isopleths were drawn to connect points of similar age. The stippled areas represent the modern range for that species.

examine this problem using the following set of equations, which is identical to eqns (6.1):

$$\frac{\partial n_1}{\partial t} = D_1 \frac{\partial^2 n_1}{\partial x^2} + (\varepsilon_1 - \mu_{11}n_1 - \mu_{12}n_2)n_1,$$

$$\frac{\partial n_2}{\partial t} = D_2 \frac{\partial^2 n_2}{\partial x^2} + (\varepsilon_2 - \mu_{21}n_1 - \mu_{22}n_2)n_2,$$

(7.1)

where we denote by $n_1(x,t)$ the population density of the earlier invading species 1, and by $n_2(x,t)$ that of the later arriving species 2. We are interested in the case when, compared with species 1, species 2 is 'weaker competitively but has a higher diffusive capability' (i.e., it is a fugitive species). Thus, the competitive relationship satisfies (iii) of (6.4), and the diffusion coefficient of species 2 is higher than that of species 1:

$$\frac{\varepsilon_1}{\mu_{11}} > \frac{\varepsilon_2}{\mu_{21}}, \qquad \frac{\varepsilon_2}{\mu_{22}} < \frac{\varepsilon_1}{\mu_{12}}, \qquad \text{and} \quad D_1 < D_2. \qquad (7.2)$$

We assume here that before species 2 is introduced, species 1 is the sole invader. Thus the population dynamics of species 1 can be expressed by letting $n_2 = 0$ in the first equation of (7.1). Because this is the same as the

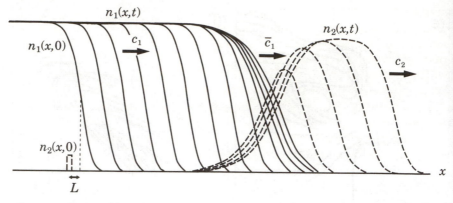

Fig. 7.2 Competition for a large open space. Invasions occur stepwise. First species 1 invades a large open area and expands its range by forming a travelling wave. Then species 2 invades a point located at distance L inside the travelling wave of species 1. Species 1 is stronger in competition but has a lower diffusive capability compared with species 2. Solid lines are spatial distributions of species 1 at set time intervals, and dashed lines are those of species 2. When species 2 invades, its density tends to decrease due to competition. However, a small number of species 2 eventually reach the range front of species 1, and then expand their range into the open space available. Parameters are chosen as $D_1 = 1$, $D_2 = 2$, $\varepsilon_1 = 1$, $\varepsilon_2 = 0.9$, $\mu_{ij} = 1$ ($i, j = 1, 2$) and $L = 3$. (After Shigesada, 1992.)

Fisher equation (3.1), species 1 is expanding its range in the form of a travelling wave of speed

$$c_1 = 2\sqrt{\varepsilon_1 D_1}. \tag{7.3}$$

Hereafter we express this travelling wave as $\tilde{n}_1(x - c_1 t)$. Next, assume that N_0 individuals of species 2 (or a mutant of species 1) invade a point located a distance L inside the travelling wave's front at time $t = 0$ (here we shall define the front as the point at which the population density near the leading edge of the travelling wave reaches half the carrying capacity). If we set the front of species 1 at time $t = 0$ to be the origin, the initial situation can be expressed as follows:

$$n_1(x, 0) = \tilde{n}_1(x),$$
$$n_2(x, 0) = N_0 \delta(x + L). \tag{7.4}$$

Solving (7.1) under these initial conditions, we can see how competition takes place. Figure 7.2 shows a numerical result. The solid and broken lines respectively indicate the range distributions of species 1 and 2, taken at set time intervals. Immediately after species 2 invades, it loses out in the competition, becoming reduced in number, while species 1 is hardly affected and continues to expand its range at speed c_1 for some time. However, a small number of species 2 eventually reach the range front of species 1, where freed from any competition species 2 dramatically recovers its population and subsequently expands its range at a constant speed into the open space available. When we used different sets of parameters from that of Fig. 7.2 for the numerical computation, we found cases in which the later invading species continued to dwindle in number and eventually became extinct.

What then are the exact conditions that make it possible for the later invading species to catch up and pass beyond the earlier species, as in Fig. 7.2? The answer to this can be obtained by examining as before the conditions for which the steady-state solution, $n_1 = \tilde{n}_1(x - c_1 t)$ and $n_2 = 0$, becomes unstable. After carrying out a stability analysis as well as numerical computation, we found two possible outcomes (see also section 7.4):

(I) if $\varepsilon_1 D_1 > \varepsilon_2 D_2$, the later invading species fails to establish itself and becomes extinct;

(II) if $\varepsilon_1 D_1 < \varepsilon_2 D_2$, the later invading species is able to catch up and pass by the earlier species, enabling it to become established, following the general pattern of Fig. 7.2.

Let us examine case (II) in more detail. As we saw above, initially species 1 is expanding at speed $c_1 = 2\sqrt{\varepsilon_1 D_1}$, but its speed falls as soon as it is passed by species 2, eventually settling at a lower constant speed. This lower speed is obtained in the same manner as eqn (6.6) of section 6.3, or

$$\bar{c}_1 = 2\sqrt{\varepsilon_1 D_1 \left(1 - \frac{\varepsilon_2 \mu_{12}}{\varepsilon_1 \mu_{22}}\right)}. \tag{7.5}$$

In other words, when two species are trying to extend their respective ranges against each other's, the stronger species 1 wins by eating away the territory of species 2 at the speed given by eqn (7.5) (Note that the numbers 1 and 2 are interchanged between eqns (6.6) and (7.5), because species 2 was the competitively stronger of the two in eqn (6.6).) Comparing c_1 and $\bar{c}_1$, we obtain

$$\bar{c}_1 = c_1 \sqrt{1 - \frac{\varepsilon_2 \mu_{12}}{\varepsilon_1 \mu_{22}}} < c_1.$$

This means that if the two species have similar growth rates and the intra- and inter-specific competition coefficients of species 2 are about the same, $\bar{c}_1$ is much smaller than c_1. This explains why, after species 1 is passed by species 2, expansion of the former's range slows down drastically. Meanwhile, once it passes species 1, species 2 is able to spread its range into open space freely, so that expansion follows the second equation of (7.1) with $n_1 = 0$ (i.e., the Fisher equation), giving a speed of

$$c_2 = 2\sqrt{\varepsilon_2 D_2}. \tag{7.6}$$

Thus, we have the expressions for c_1, $\bar{c}_1$ and c_2 in eqns (7.3), (7.5) and (7.6), respectively. When these values are calculated using the parameters of Fig. 7.2, we have $c_1 = 2$, $c_2 = 2.68$ and $\bar{c}_1 = 0.632$; and so

$$\bar{c}_1 \ll c_1 < c_2. \tag{7.7}$$

Therefore, once species 2 passes species 1, it will continue to widen the gap to move into and occupy the available open space.

The condition $\varepsilon_1 D_1 < \varepsilon_2 D_2$ of case (II) can be rewritten using eqns (7.3) and (7.6) as $c_1 < c_2$. This means that when there is no competition, the rate of range expansion is higher for species 2 than for species 1, and it is only under this condition that species 2 is able to slip by species 1's range and jump ahead. In other words, the species which has a higher value for the product εD of the diffusion coefficient and growth rate is the one that reaches and occupies open space faster eventually; even a species with higher diffusive capability will fail in invading if it has a low growth rate. Note, however, that if it must endure a long period of extremely low population before reaching the earlier species' range front, even a species with a higher εD value may become extinct due to demographic stochasticity (see section 2.5).

Assuming that it does not become extinct in the process, we can calculate the time t^* required for the late species to pass by the earlier one (see section 7.4). That is, if at $t = 0$, N_0 individuals invade a point located a distance L inside the travelling wave front of the earlier species, t^* is the time at which the descendants of the later species reach the earlier species' front and exceed a threshold density n^*. Figure 7.3 shows that t^* increases roughly linearly with L, the distance from wave front to invasion site. Moreover, t^* is affected little by the threshold density n^* or the initial invading population N_0 (see the discussion in section 7.4).

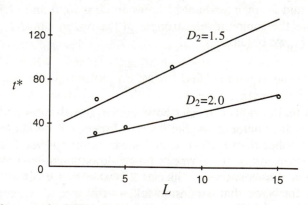

Fig. 7.3 Time required for a late-coming species to pass by an earlier one as a function of the invasion site, L, for $D_2 = 1.5$ and 2.0. See section 7.4 for details.

(a)

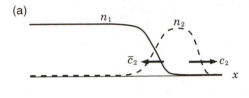

(b)

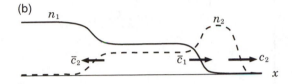

(c)

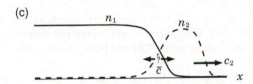

Fig. 7.4 Possible outcomes of the competition for space when $\varepsilon_1 D_1 < \varepsilon_2 D_2$. Invasions of two species occur stepwise as in Fig. 7.2. When $\varepsilon_1 D_1 < \varepsilon_2 D_2$ holds, species 2 eventually reaches the range front of species 1. The spatial patterns of the two species asymptotically attained are classified into four cases depending on whether competition in the absence of diffusion leads to (i) extinction of species 1, (ii) coexistence, (iii) extinction of species 2, or (iv) contingent extinction of either species. The results for cases (i), (ii), and (iv) are shown in (a), (b) and (c), respectively. Case (iii) has already been illustrated in Fig. 7.2. For the speed $\bar{c}$ in (c) see Fig. 7.5.

Although we have only considered the case that the later invading species is competitively inferior to the earlier species, the main result above, namely that the species with a higher value of εD will reach open space faster and expand its range in the form of a travelling wave, remains true for the other three cases of competitive relations as given by (i), (ii) and (iv) in (6.4). As seen in section 6.3, however, the rear of the frontal wave differs depending on the three cases. Figure 7.4 illustrates the typical spatial patterns of two species that are asymptotically attained after species 2 passes by species 1. When species 2 is competitively stronger (i.e., (6.4(i)) holds), the boundary of the two species moves to the left, replacing the earlier species by the later species at velocity

$$\bar{c}_2 = 2\sqrt{\varepsilon_2 D_2 (1 - \varepsilon_1 \mu_{21}/\varepsilon_2 \mu_{11})}$$

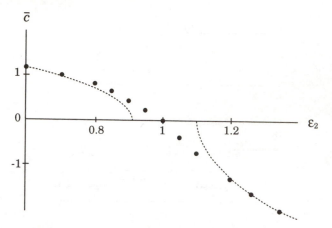

Fig. 7.5 Speed of boundary, $\bar{c}$, against ε_2 when the outcome of the competition is contingent. Other parameters are fixed as $D_1 = 1$, $D_2 = 4$, $\varepsilon_1 = 1$, $\mu_{11} = \mu_{22} = 1$ and $\mu_{12} = \mu_{21} = 1.1$. When $\varepsilon_2 < 1$, the boundary moves to the right, and when $\varepsilon_2 > 1$, it moves to the left. The dotted curves on the left and right sides represent approximate solutions given by eqns (7.5) and (6.6), respectively.

as given by eqn (6.6). When the two species coexist in the absence of dispersal (i.e., (6.4)(ii)) holds), the left and right fronts of the coexisting region expand at speeds

$$\bar{c}_2 = 2\sqrt{\varepsilon_2 D_2 (1 - \varepsilon_1 \mu_{21} / \varepsilon_2 \mu_{11})}$$

and

$$\bar{c}_1 = 2\sqrt{\varepsilon_1 D_1 (1 - \varepsilon_2 \mu_{12} / \varepsilon_1 \mu_{22})},$$

respectively. When the outcome of competition is contingent (i.e., (6.4)(iv)) holds), it has been shown that whether the boundary of the two species' ranges moves to the right or left is determined by the parameter values (Namba and Mimura, 1980; Kan-on, 1995). For example, Fig. 7.5 shows how fast and which direction the boundary moves depending on the value of ε_2.

7.3 Disturbance and species diversity

Although the above discussion assumed on infinite open space, in real situations there is a limit to the amount of open space available. It would seem that a competitively weaker species, even if it has a larger εD, is doomed to extinction because eventually the stronger species is bound to catch up. For the competitively weaker species to survive, therefore, a fresh supply of open space must be made available by disturbances etc. on

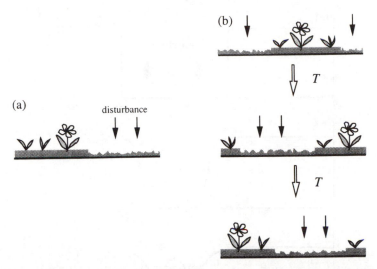

Fig. 7.6 Two types of disturbance. Disturbance occurs at constant intervals of T. At each disturbance, an open space is created (a) at the right half of the area, or (b) at a randomly selected area which occupies half of the total area. Right and left boundaries are connected to each other (i.e., periodic boundary condition).

a regular basis. Here, we use eqns (7.1) to simulate competition in a finite environment where an open space of fixed size is periodically generated. For simplicity, we assume that the range is one-dimensional and satisfies a periodic boundary condition (i.e., the population densities at both ends are the same). As before, species 1 is competitively stronger than, but inferior in diffusive capability to, species 2, that is, the conditions (7.2) hold. The disturbance occurs at constant intervals of T, turning half of the total area into open space, from which the resident organisms are completely wiped out. Since $1/T$ is the frequency at which the disturbance occurs per unit time, the larger $1/T$ is, the harsher the environment. We consider two ways in which open space is created (see Fig. 7.6):

(1) regular disturbance—dividing the range at its centre, only the right half is subjected to disturbances to become an open space;
(2) random disturbance—the disturbance occurs at a randomly selected location.

We first took case (1), where the disturbance occurs regularly at the same place, and numerically computed eqns (7.1) to see how the two species' distribution varies with time. An example is shown in Fig. 7.7, with $\varepsilon_1 = 1$, $\varepsilon_2 = 0.9$, $\mu_{ij} = 1$ $(i, j = 1, 2)$, $D_1 = 1$, and $D_2 = 4$. After the initial transient state, the two species settle to a temporary periodic distribution as shown in Fig. 7.7(a). Using these distribution diagrams, we calculated

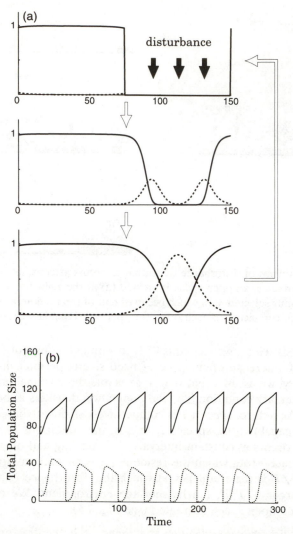

Fig. 7.7 Competition for open space under regular disturbances. (a) Change in the spatial patterns of two competing species. Solid and dotted curves represent distributions of species 1 (competitively stronger species) and species 2 (fugitive species), respectively. (b) Change in total population sizes with time. Parameters are chosen as $D_1 = 1$, $D_2 = 4$, $\varepsilon_1 = 1$, $\varepsilon_2 = 0.9$, $\mu_{ij} = 1$ and $1/T = 0.03$ ($T = 33$).

the total population sizes of the two species over the entire range, which are plotted in Fig. 7.7(b). The population sizes drop suddenly following each disturbance, but then continue to recover until the next disturbance; this pattern is repeated, eventually settling to a stable, periodic oscillation

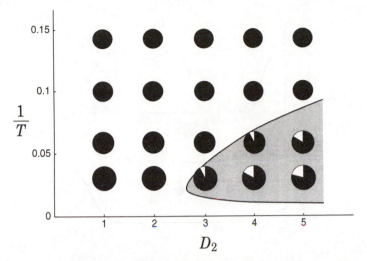

Fig. 7.8 Average total population sizes of two competing species under regular disturbances for varying $1/T$ and D_2. Vertical axis, $1/T$, represents the frequency of disturbance. Horizontal axis represents the diffusion coefficient of the fugitive species (species 2). Other parameters are chosen as in Fig. 7.7. The black and white areas in each circle indicate time-averaged population sizes of species 1 and species 2, respectively. Species 1 can always survive, while species 2 is allowed to persist only when D_2 is sufficiently large, or when the frequency of disturbance is intermediate (stippled area). (After Shigesada, 1992)

where the two species coexist. When the numerical computation was carried out using different sets of T and D_2, we found that there were instances where species 2 became extinct with only species 1 surviving. Figure 7.8 shows the results plotted in the $(D_2, 1/T)$ parameter plane. Thus we see that while species 1 invariably survives, species 2 on the other hand survives only when its diffusion coefficient D_2 is large and the disturbance occurs with intermediate frequencies. In other words, the weaker species 2 persists only by quickly reaching an open space to recover its population (this requires that $\varepsilon_1 D_1 < \varepsilon_2 D_2$ as we saw in section 7.2), and then moving to another open space created by the next disturbance before the stronger species arrives. But the weaker species can never displace the stronger one to become the sole surviving species. The stronger species 1 is never threatened by extinction even if its migrating speed is slow, because as seen from Fig. 7.7(a), when an open space is newly created by a disturbance, there is always a large number of species 1 present at adjacent locations, enabling it to move in immediately.

We next did a similar analysis for case (2), when the open space is created at a randomly selected site. Figure 7.9 shows how the spatial

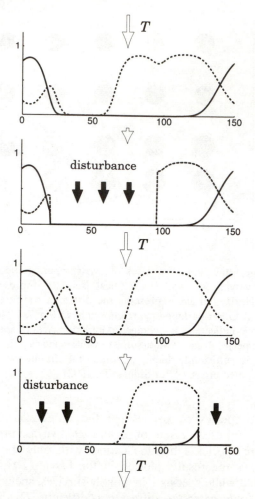

Fig. 7.9 Change in spatial patterns of two competing species under random disturbances. Solid and dotted curves represent distributions of species 1 and species 2, respectively.

distributions of the two species vary with time, while Fig. 7.10 illustrates the change in the total population sizes of two species with time. Not surprisingly, neither of them settles to a periodic pattern. From Fig. 7.10, we see that in addition to cases where (a) only the stronger species survives and (b) the two species coexist, there are instances where (c) only the weaker species survives. Summarizing the results on the $(D_2, 1/T)$ parameter plane, as we did earlier in Fig. 7.8, we obtain Fig. 7.11. The zone in which the two species coexist corresponds to disturbances occurring with

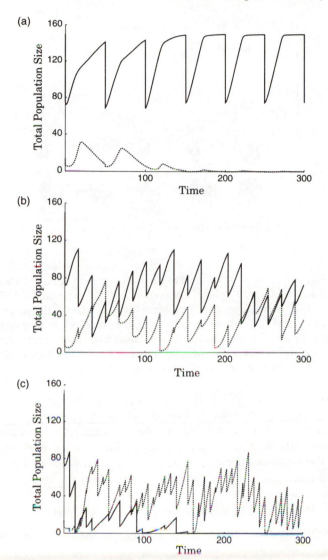

Fig. 7.10 Change in total population sizes of two competing species with time under random disturbances: (a) When $1/T = 0.02$ ($T = 50$), species 2 becomes extinct; (b) when $1/T = 0.06$ ($T = 17$), both species can coexist; (c) when $1/T = 0.14$ ($T = 7$), species 1 becomes extinct. All parameters except T are the same as in Fig. 7.7.

intermediate frequencies as before, although it occupies a more extensive area than in Fig. 7.8. The zone representing cases when only the weaker species 2 survives lies in the region where $1/T$ is large (i.e., disturbance is

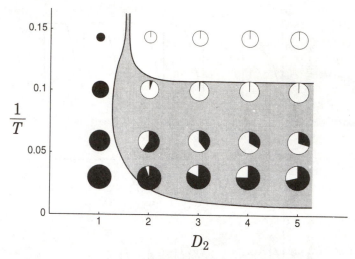

Fig. 7.11 Average total population sizes of two competing species under random disturbances in the $(D_2, 1/T)$ plane. Parameters are the same as in Fig. 7.8. When either D_2 or $1/T$ is small, species 1 always wins in the competition, while in the region where D_2 and $1/T$ are sufficiently large, only species 2 is allowed to persist. When the frequency of disturbance $1/T$ is intermediate, both species can coexist. The parameter region where coexistence occurs is larger than that in the case of regular disturbance (compare stippled area with that in Fig. 7.8).

harsh) and D_2 is large. Note that while the stronger species never faces extinction under regularly located disturbances, it becomes extinct under certain conditions for randomly located disturbances; this is because in the latter instance there is the chance that an area with a high population density of the stronger species can be hit by a disturbance and turned into an open space, causing great damage to the stronger species with slow movement.

Summarizing cases (1) and (2) above, we can state that

 (i) intermediate disturbances, whether they are regular or random ones, promote coexistence;
 (ii) under regular disturbances, the competitively stronger species has the survival advantage;
(iii) under random disturbances, the competitively weaker species is able to adapt and, with a high diffusive capability, can at times even displace its competitor to become the sole surviving species.

Conclusion (i) above provides theoretical backing to Connell's assertion mentioned at the beginning, namely that 'intermediate disturbances promote coexistence among many species and thus increase an ecosystem's biodiversity'. We can also conclude from above that a system's biodiversity increases if the sites of disturbance are more randomly located.

The effects of disturbance on the invasion of a competitively inferior species into communities consisting of more than two competing species were examined by Rejmanek (1984, 1989) by using an extended version of eqns (7.1) for multiple species. Although the disturbances he adopted were different in size and strength from those in the above model, his results also confirm the intermediate disturbance hypothesis.

Although we have used diffusion-reaction equations to describe the spatio-temporal pattern of populations, there are potentially many other modelling approaches, depending on whether space, time and population density are continuous or discrete variables (Caswell and Etter, 1992). To explore how intermediate disturbance is advantageous to a fugitive species, Etter and Caswell (1994) proposed a cellular automaton model, in which space, time and population density are all discretized: two-dimensional space is cut into a regular array of cells (patches), in which two species (S_1 and S_2; S_2 is fugitive) are competing for space. Each cell is characterized by a finite number of states as empty or occupied by S_1, S_2 or both, and when S_1 and S_2 co-occur, S_1 always excludes S_2. S_1 is allowed to colonize immediately adjacent cells, while S_2 can colonize any empty patch that is produced by disturbance. Using a cellular automaton consisting of 256×256 cells, Etter and Caswell performed a simulation to show that the advantage to fugitive species peaks at intermediate disturbance frequencies and the disturbance frequency at which it peaks increases with the rate of competitive exclusion.

The succession process of disturbance-mediated systems was studied by Levin and Buttel (1987) by using an interacting particle model, in which both space and population density are discretized, while time is continuous. Each cell was characterized by one of a small number of successional states. Local states could be altered due to invasion by a species with later successional characteristics, or by disturbance, which reset cells to the initial state. This model was applied to four annual plants in serpentine grasslands where disturbances caused by gophers influence the dynamics of plants (see also Levin *et al.*, 1989; Moloney *et al.*, 1992). Their computer simulations showed that some disturbance regions lead to the extinction of particular species, while others maintain coexistence through ergodic spatio-temporal mosaics, so that the timing and magnitude of disturbance are both important (see also Levin, 1992).

Although we have introduced only a few studies, there are many other approaches to modelling the dynamics of disturbance-mediated spatially distributed systems (see Pacala and Silander, 1985; Colasanti and Grime, 1993; Durrett and Levin, 1994; Harada and Iwasa, 1994; Harada *et al.*, 1995; Kubo *et al.*, 1996). If we ignore the spatial arrangement of patches (individuals migrate to all patches equally), the problem can be treated by a classical metapopulation model, which also shows the capability of a fugitive species to coexist with a competitively stronger species (Slatkin,

1974; Levin and Paine, 1974; Armstrong, 1976; Caswell, 1978; Hastings, 1980; Hanski, 1983; Caswell and Cohen, 1991; see also the book edited by Gilpin and Hanski, 1991). As another line of approach, the effects of disturbance on species diversity were studied by using multispecies Lotka–Volterra systems, where every individual is assumed to interact equally with every other individual, by Huston (1979), Hastings (1980) and Teramoto (1993); see also the review by DeAngelis *et al.* (1985).

7.4 Appendix: range expansion of a later invading species: an approximate solution

As described in section 7.2, we consider a situation where first species 1 is expanding in the form of a travelling wave $\tilde{n}_1(x - c_1 t)$ and then some propagules of species 2 invade a point located a distance L inside the front of species 1. If the initial density of species 2 is small, its time development can be examined by the following linearized equation of (7.1) for some early period:

$$\frac{\partial n_2}{\partial t} = D_2 \frac{\partial^2 n_2}{\partial x^2} + (\varepsilon_2 - \mu_{21}\tilde{n}_1(x - c_1 t))n_2, \qquad (7.8)$$

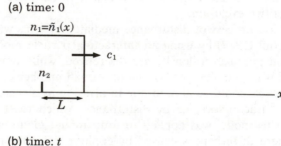

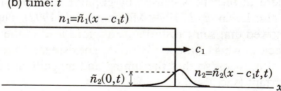

Fig. 7.12 Invasion of species 2 into an open space where species 1 has been spreading in a travelling wave with a step function. (a) Species 2 invades a point at distance L behind the front of species 1 at $t = 0$. (b) If the travelling wave of species 1 moves to the right at speed c_1 without changing its shape, the spatial distribution of species 2, $n_2(x, t) = \tilde{n}_2(x - c_1 t, t)$, and its density at the front of species 1, $n_2(c_1 t, t) = \tilde{n}_2(0, t)$, are exactly given by eqns (7.11) and (7.12), respectively, as described in the text.

with the initial condition,

$$n_2(x,0) = N_0 \delta(x+L). \qquad (7.9)$$

Since the travelling wave of species 1, $\tilde{n}_1(x - c_1 t)$, has a sigmoidal shape with ε_1/μ_{11} at $x \to -\infty$ and 0 at $x \to \infty$, we further assume that $\tilde{n}_1(x - c_1 t)$ is approximated by the following step function (see Fig. 7.12):

$$\tilde{n}_1(x - c_1 t) = \begin{cases} \dfrac{\varepsilon_1}{\mu_{11}} & \text{for } x - c_1 t < 0 \\ 0 & \text{for } x - c_1 t > 0 \end{cases}, \qquad (7.10)$$

By using the method of Green's functions (Carslaw and Jaeger, 1959; p. 357), eqn (7.8) with eqns (7.9) and (7.10) can be solved as follows:

$$n_2(x,t) = \tilde{n}_2(z,t)$$

$$= \begin{cases} A(z,t)\left[\dfrac{1}{\sqrt{4\pi D_2 t}} \left\{ \exp\left(-\dfrac{(z+L)^2}{4D_2 t} \right) - \exp\left(-\dfrac{(z-L)^2}{4D_2 t} \right) \right\} - B(z,t,0) \right], \\ \hspace{8cm} (z < 0) \\ A(z,t)B(z,t,v), \\ \hspace{8cm} (z > 0) \end{cases}$$

$$(7.11)$$

and

$$n_2(ct,t) = \tilde{n}_2(0,t) = A(0,t)\phi(t), \qquad (7.12)$$

where

$$z = x - c_1 t, \qquad v = \varepsilon_1 \mu_{21}/\mu_{11},$$

$$A(z,t) = N_0 \exp\left\{ -\dfrac{c_1}{2D_2} z + (\varepsilon_2 D_2 - \varepsilon_1 D_1)\dfrac{t}{D_2} - vt \right\},$$

$$\phi(t) = \dfrac{1}{\pi v \sqrt{D_2}} \left[\int_0^\infty (\sqrt{\rho+v} - \sqrt{\rho}) e^{-\rho t} \cos\left(L\sqrt{\rho/D_2} \right) d\rho \right.$$

$$\left. + \int_0^v \sqrt{v-\rho}\, e^{\rho t - L\sqrt{\rho/D_2}}\, d\rho \right]$$

$$B(z,t,v) = \dfrac{z}{\sqrt{4\pi D_2}} \int_0^t \phi(\tau) \dfrac{\exp\left\{ v(t-\tau) - \dfrac{z^2}{4D_2(t-\tau)} \right\}}{(t-\tau)^{3/2}}\, d\tau.$$

Here $\tilde{n}_2(z,t)$ represents the spatial pattern of species 2 if we look at it in a travelling frame that is moving to the right at speed c_1, and $\tilde{n}_2(0,t)$ is the

$\tilde{n}_2(0, t)$

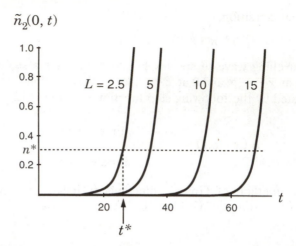

Fig. 7.13 Density of species 2 at the front of species 1, $\tilde{n}_2(0,t)$, as a function of time for varying L. Parameters are $c_1 = 2$, $D_2 = 2$, $\varepsilon_1 = \varepsilon_2 = 1$, $\mu_{11} = \mu_{21} = 1$. Dotted line indicates the threshold density; t^* is the time for $\tilde{n}_2(0,t)$ to reach n^*. See Fig. 7.3 for t^* plotted as a function of L.

population density of species 2 at the front of the travelling wave of species 1. We can see from eqn (7.12) that $\tilde{n}_2(0,t)$ increases with time when $\varepsilon_1 D_1 < \varepsilon_2 D_2$, while it eventually goes to zero when $\varepsilon_1 D_1 > \varepsilon_2 D_2$. Figure 7.13 shows how $\tilde{n}_2(0,t)$ grows with time when $\varepsilon_1 D_1 < \varepsilon_2 D_2$. As defined earlier, we calculate t^* as the time at which $\tilde{n}_2(0,t)$ reaches a certain threshold value (dotted line); t^* thus obtained is plotted in Fig. 7.3. We can see from Fig. 7.13 that t^* changes slightly if we take different threshold values, because $\tilde{n}_2(0,t)$ starts at zero, stays at a very low level for a while and then sharply increases afterwards.

8

Invasion of predators

8.1 Invasion of a predator and biological control of pest species

Predators include carnivores such as lions and tigers, herbivores such as zebras and elephants, and omnivores such as humans and apes. Predation exists not only among mammals, but in all levels of species including birds, fishes, insects, bacteria, etc. As we saw in Chapter 2, because of the extensive damage to agriculture, cases of invasion by herbivorous insects have been well documented. Parasites or parasitoids which lay eggs in the larvae of such plant-eating pests can also be considered as predators.

Insect pests can be biologically controlled by introducing parasitoids which are their natural enemies. Although massive outbreaks of insect pests are rarely seen in their native land, when an insect pest is introduced to a new location it is usually unaccompanied by its natural enemy, and thus is often able to multiply without control (Pimentel, 1986). A successful example of pest control by introducing its natural enemy is seen in the control of the *Icerya purchasi* outbreak in California. This species, which causes damage to citrus trees, was brought under control by introducing in 1888 its natural enemy, *Rodolia cardinalis*. Worldwide, there are over 100 completely successful cases of pest control achieved through the introduction of a natural enemy; this list grows to about 200 cases when partial successes are included as well. Hassell and May have carried out extensive studies on mathematical models of biological control by the introduction of natural-enemy species; they have examined for various systems the conditions for initial establishment of a natural-enemy species, for maximum depression of the host population and for persistence of the populations in a stable interaction, etc. See May and Hassell (1988) for a detailed review.

Another biological control method to prevent the spread of an invading species is the release of sterile insects. An example is seen in the control of the melon fly (*Dacus cucurbilae*), which was first introduced to Yaeyama Islands of Okinawa, Japan, in 1919 (Iwahashi, 1977): vast numbers of male insects whose reproductive cells were destroyed by gamma irradiation were released; this prevented the females from laying normal eggs, and thus led to their successful extermination in 1993. Although the programme entailed large investments for the construction of irradiation facilities, its success made it possible to lift the ban on transporting melons, thus bringing about considerable economic benefits.

As we have seen above, studies on the conditions whereby a predator (or parasitoid) species succeeds in initial establishment as well as on its use in biological control have received considerable attention because of their potential practical benefits, and many noteworthy findings have come out of this field. Below, we introduce three studies which dealt with the spatio-temporal dynamics of predator-prey systems. In the following section, we discuss a study of Dunbar (1983) on the invasion conditions of a predator species and the expansion rate of its range, using the Lotka–Volterra predator–prey model with limited growth of prey. Next, we introduce the works of Hassell *et al.* (1991) and Comins *et al.* (1992) on self-organized spatial structures of insect host-parasitoid systems. Finally, we look at the mathematical model of Lewis and van den Driessche (1993) that examines how the release of sterile insects controls the invasion speed of an insect pest.

8.2 Travelling wave in prey–predator system

We consider the invasion of a predator into an environment which a prey species is inhabiting homogeneously in one-dimensional space. We use the Lotka–Volterra model for the interaction between predator and prey, both of which are dispersing by diffusion. With the prey density and predator density denoted by h and p, respectively, the equations for their rates of change are given by

$$\frac{\partial}{\partial t} h = D_1 \frac{\partial^2}{\partial x^2} h + \varepsilon \left(1 - \frac{h}{K}\right) h - ahp, \qquad (8.1a)$$

$$\frac{\partial}{\partial t} p = D_2 \frac{\partial^2}{\partial x^2} p - \delta p + bhp, \qquad (8.1b)$$

where the first term of the right-hand side of both equations indicates diffusion. The second term in eqn (8.1a) represents the logistic growth rate of prey in the absence of predators and the third one is the consumption rate of prey by predators. The second term on the right-hand side of eqn

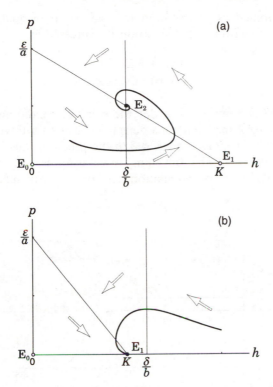

Fig. 8.1 Phase plane diagrams of the Lotka–Volterra prey–predator system. (a) When $K > \delta/b$, both prey and predator can coexist. (b) When $K < \delta/b$, the predator fails in invasion and only the prey persists, keeping its density at its carrying capacity, K.

(8.1b) is the death rate of predators and the last one is the rate of increase in the number of predators from ingesting prey.

We first examine how the population densities change with time when the diffusion terms are ignored in eqns (8.1) (i.e., when $D_1 = D_2 = 0$). By conducting a phase plane diagram as shown in Fig. 8.1, we can see that there are three equilibria $E_0(0,0)$, $E_1(K,0)$ and $E_2(\delta/b, \varepsilon(bK - \delta)/abK)$. When $K > \delta/b$, E_2 is positive and dynamically stable (that is, stable coexistence of both species), while the other equilibria are unstable. When $K \leq \delta/b$, E_2 disappears from the first orthant and E_1 becomes stable, that is, the predator cannot persist and only the prey survives, keeping its density at the carrying capacity, K. Thus invasion of the predator succeeds only if the prey's abundance (its carrying capacity) exceeds δ/b.

Next we examine eqns (8.1) with diffusion terms. Assume that the prey density has already attained the carrying capacity K uniformly in space, when some propagules of the predator (say n_0 individuals) invade the origin at $t = 0$. Then the initial situation is expressed as

$$h(x,0) = K,$$
$$p(x,0) = n_0 \delta(x). \tag{8.2}$$

Solving eqns (8.1) numerically under these initial conditions, we can see that when $K > \delta/b$, the predator's density evolves to a travelling wave as shown in Fig. 8.2, whereas when $K \leq \delta/b$ the predator fails in its invasion. The spatial pattern of the travelling wave exhibits two distinct types: behind the wave front, the population densities of both species approach

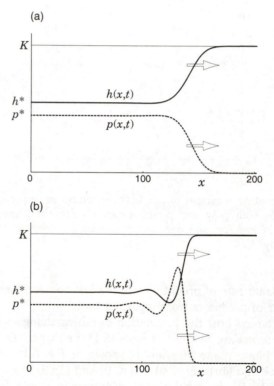

Fig. 8.2 Travelling wave solutions of eqns (8.1) (modified from Dunbar, 1983). As the initial condition, a few propagules of the predator locally invade an area where the prey exists keeping its density at its carrying capacity, K. When $K > \delta/b$, a travelling wave connecting two equilibrium states, $(K,0)$ and $(h^*, p^*) = (\delta/b, \varepsilon(bK - \delta)/abK)$, always evolves. Parameters are chosen as (a) $b = 0.6$, $\delta = 0.33$, for which the travelling wave has a monotonically damping tail, and (b) $b = 4.0$, $\delta = 2.2$, for which the travelling wave has a damped oscillating tail. Other parameters are $D_1 = D_2 = 1$, $\varepsilon = 1$, $a = 1$, $K = 1$.

equilibrium E_2 either (a) monotonically or (b) with damping oscillation. The mathematical proof of the existence of such travelling waves was presented by Dunbar (1983 and 1984) (see also Murray (1989) for the case $D_2 = 0$). He showed that the speed of the travelling wave, c, should satisfy

$$c \geq c_{min} = 2\sqrt{D_2(-\delta + bK)}. \qquad (8.3)$$

In analogy with results from the Fisher equation (see section 3.9), he conjectured that the stable speed actually realized is the minimum one, c_{min}. Thus the rate of range expansion of predator increases with the prey's carrying capacity K if $K > \delta/b$. If $K < \delta/b$, however, the travelling wave never evolves, resulting in the predator's extinction, as expected from the diffusion-free model.

Here we heuristically derive eqn (8.3) in a way similar to that used to obtain eqn (6.6) for a competition system. At the wave front of the predator, the population densities of prey (h) and predator (p) are approximately K and 0, respectively, and so eqn (8.1b) is approximated as

$$\frac{\partial}{\partial t} p = D_2 \frac{\partial^2}{\partial x^2} p + (-\delta + bK)p. \qquad (8.4)$$

Since this has the same form as the Skellam equation, the speed of the travelling wave is given by c_{min} in eqn (8.3).

Later, in Chapters 9 and 10, the present model will be extended and applied to the spreading of epidemic diseases.

A prey–predator system similar to eqns (8.1) can also explain the geographical spread of early agriculture from the Middle East into Europe, which was then occupied by indigenous hunter-gatherers (Aoki *et al.*, 1996). Upon encounter with a farmer, a hunter-gatherer will be converted to a farmer with a certain probability. This interaction can be formally described by the same mass action term used in a prey and predator system.

8.3 Host–parasitoid systems and self-organized spatial structures

Parasitoids are insects which lay their eggs in the larvae or pupae of arthropod hosts. The larval parasitoids kill the host as they feed on it. Host–parasitoid systems generally involve non-overlapping generations, with reproduction taking place at regular intervals particularly in temperate regions. If parasitoids search independently of each other and encounter hosts at random, the changes in their population sizes are described by the following difference equations, which were originally presented by Nicholson and Bailey (1935):

$$h_{t+1} = \lambda h_t \exp(-ap_t),$$
$$p_{t+1} = ch_t(1 - \exp(-ap_t)), \tag{8.5}$$

where h_t and p_t are the number of female hosts and parasitoids in generation t, respectively, λ is the host intrinsic reproductive rate, c is the number of parasitoids emerging from a parasitized host which survive to next generation, and $\exp(-ap_t)$ is the probability that a host individual escapes parasitism, where a is called the effective area of search and represents a measure of the parasitoid searching efficiency. Since we have assumed that parasitoids search independently and randomly, the number of parasitoids which discover a particular host follows a Poisson distribution with mean ap_t; $\exp(-ap_t)$ is then the zeroth term of the Poisson series, representing the probability that no parasitoid discovers the host.

Equations (8.5) have two equilibria,

$$(0,0) \quad \text{and} \quad (\lambda \log \lambda / ac(\lambda - 1), \log \lambda / a),$$

both of which are shown from linear stability analysis to be always unstable. Computer simulations also demonstrate that the solution is globally unstable with ever-diverging oscillations, resulting in extinction first of the hosts and consequently of the parasitoids due to demographic stochasticity.

To stabilize the host–parasitoid association, various refinements have been introduced to the interaction between host and parasitoids. Thus, Beddington *et al.* (1975) presented an extended version of the Nicholson–Bailey model which incorporates the density effect on the host population's growth rate by substituting λ for $\lambda \exp(-\gamma h_t)$ (where γ is a positive constant) in eqn (8.5). This model generates a stable equilibrium, stable limit cycles and even chaos, depending on the parameter values. In particular, the limit cycles exhibit large oscillatory amplitudes that are qualitatively similar to those observed in an experimental system of the parasitoid braconid wasp and its host the azuki bean weevil by Uchida (1957) (see Brown and Rothery, 1993). May and Hassell (1988) and May (1994b), however, have pointed out that such interactions that include the density effects of hosts or mutual interference among searching parasitoids are not generally influential enough to explain the persistence of host–parasitoid associations in the natural world. They suggest rather that spatial heterogeneity coupled with sufficient variation in the parasitoid attack rates among patches is the essential mechanism that enables the persistence of host and parasitoid populations in field situations (see also Pacala *et al.*, 1990).

Meanwhile, Hassell *et al.* (1991) and Comins *et al.* (1992) carried out computer simulations of the Nicholson and Bailey model in a spatial

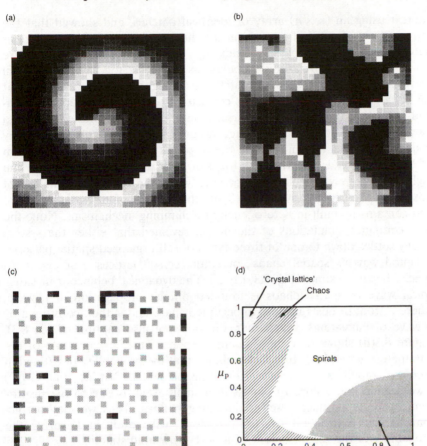

Fig. 8.3 Instantaneous maps of population density for simulations of the local dispersal model with Nicholson–Bailey local dynamics (after Comins *et al.*, 1992). Black squares represent empty patches; dark shades becoming paler represent patches with increasing host densities; light shades to white represent patches with hosts and increasing parasitoid densities. (a) $\mu_h = 1$, $\mu_p = 0.89$, spiral structure: local population densities form spiral waves which rotate in either direction around almost immobile focal points. (b) $\mu_h = 0.2$, $\mu_p = 0.89$, spatial chaos: host and parasitoid populations fluctuate from patch to patch with no long-term spatial organization. (c) $\mu_h = 0.05$, $\mu_p = 1$, crystalline structure: patches of relatively high densities occur mostly at a spacing of two grid units and this pattern is statistically stable. (d) Dependence of observed persistent spatial pattern on μ_h and μ_p. Other parameter values are fixed as $\lambda = 2$ and $n = 30$. Shaded area represents parameter combinations for which the persistent spatial pattern is unlikely to be established by starting the simulation with a single non-empty patch; spiral may be established in these cases by starting with a lower μ_h and increasing it after 50–100 generations.

context using an $(n \times n)$ array of identical patches, and showed that the introduction of local dispersal of the host and parasitoid populations between neighbouring patches leads to a self-organized spatial structure which enable host–parasitoid populations to persist together, even when there is no variability between patches (see Fig. 8.3). They assumed that for each generation, the dynamics consists of two phases: a reproduction-and-parasitism phase and a dispersal phase. In the first phase, hosts and parasitoids in each patch interact according to the Nicholson–Bailey dynamics as given by eqn (8.5). In the dispersal phase, some fixed fraction of hosts (μ_h) and of parasitoids (μ_p) in each patch move evenly to the eight immediately neighbouring patches while the remaining fraction of hosts and parasitoids stays home. Thus this model is purely deterministic and contains no built-in heterogeneity or clumping mechanisms. None the less, computer simulations of the model revealed that either the system finally settles down to one of three types of self-organized spatial patterns —'spiral waves', 'spatial chaos', or static 'crystal lattices'—or else both species become extinct (Fig. 8.3(a, b, c)). The dynamical behaviour in either spiral wave or spatial chaos is founded on the propagation of a wave, where a front of hosts invades unoccupied space and is then consumed by a wave of parasitoids, which then die out (see also caption of Fig 8.3). Figure 8.3(d) shows how the outcome depends on the magnitude of the parameters μ_h and μ_p. It was also shown that the probability of extinction increases rapidly as the arena size n falls below 30. Comins *et al.* (1992) have argued that failure to persist in a small arena is due to the insufficient space to produce a self-maintaining pattern. Thus we may predict that habitat destruction or patch removal will result in lowering the probability of the establishment of a host–parasitoid system.

Nowak *et al.* (1995) have extended the above approach to evolutionary games such as the Prisoner's Dilemma and found that if the game is played with neighbours in a two-dimensional array, cooperative behavior can evolve to generate self-organized spatial structures.

8.4 Retreat of travelling wave by sterile insect release

A female that mates with a sterile male cannot lay normal eggs. Sterile males can be produced by artificially breeding the targeted insect and exposing it to radiation. When vast numbers of these male insects are released in the field, healthy, normal females will mate with the sterile males and become unable to lay normal eggs. This method of pest control is called the sterile insect release method (SIRM), and has been successfully employed against various insect species including the screwworm fly (Baumhover *et al.*, 1955), fruit fly (Ito, 1977; Iwahashi, 1977), codling moth and pink bollworm. As the number of field applications of this method has grown, several mathematical models of SIRM have been announced. Here we introduce the work of Lewis and van den Driessche (1993), which examines how the release of sterile insects controls the invasion speed of an insect pest in the process of range expansion.

Consider an insect that disperses randomly by means of simple one-dimensional diffusion and whose birth and death dynamics is governed by the SIRM model presented by Barclay and Mackauer (1980). In this model, it is assumed that population growth follows a logistic curve; mating is at random and the frequency of fertile matings is in proportion to the fractional density of fertile individuals; the population sex ratio is 1 to 1 throughout life, and sterile individuals are released at a constant rate per unit of habitat. Thus, denoting by $u(x, t)$ and $n(x, t)$ the densities of fertile females and sterile females, respectively, we have the following dynamics for both species:

$$\frac{\partial u}{\partial t} = D_1 \frac{\partial^2 u}{\partial x^2} + \left(\frac{a_1 u}{u + n} - a_2 \right) u - 2\gamma(u + n)u,$$

$$\frac{\partial n}{\partial t} = D_2 \frac{\partial^2 n}{\partial x^2} + r - a_2 n - 2\gamma(u + n)n,$$

(8.6)

where a_1 is the birth rate, a_2 is the density-independent death rate, γ is the density-dependent death rate, r is the release rate of sterile insects, and D_1 and D_2 are diffusion constants for fertile and sterile insects, respectively. To simplify analysis, we non-dimensionalize eqns (8.6) by choosing

$$x^* = x \sqrt{a_1/D_1}, \qquad t^* = a_1 t, \qquad u^* = \frac{\gamma}{a_1} u, \qquad n^* = \frac{\gamma}{a_1} n,$$

$$A = \frac{a_2}{a_1}, \qquad R = \frac{\gamma r}{a_1^2}, \qquad \delta = \frac{D_2}{D_1}.$$

(8.7)

Thus eqns (8.6) are rewritten as

$$\frac{\partial u}{\partial t} = \frac{\partial^2 u}{\partial x^2} + \left\{ \frac{u}{u + n} - A - 2(u + n) \right\} u,$$

(8.8a)

$$\frac{\partial n}{\partial t} = \delta \frac{\partial^2 n}{\partial x^2} + R - An - 2(u + n)n,$$

(8.8b)

where we have dropped the asterisks for notational simplicity. Thus the model now includes only three dimensionless parameters, A, R and δ.

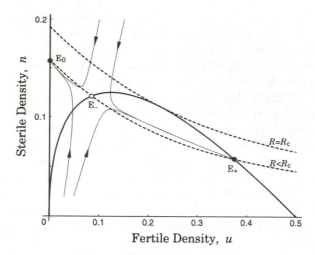

Fig. 8.4 Phase plane diagram of eqns (8.8) for the case $A = 0$ and $D_1 = D_2 = 0$. Solid line, null cline of eqn (8.8a); dotted lines, null clines of eqn (8.8b) for $R = 0.05$ and $R = R_c = 0.0741$. Arrows show typical trajectories for $R = 0.05$.

Lewis and van den Driessche estimated $A = 0$–0.3 as a biologically realistic value. δ is smaller than 1 because irradiation generally reduces the dispersal ability of insects. R is varied through manipulating the release rate r.

If no sterile individual is released (i.e., $n = 0$), eqns (8.6) or (8.8) are reduced to the Fisher equation. Thus the fertile insects locally introduced always evolve into a travelling wave that advances at a constant speed, $2\sqrt{D_1(a_1 - a_2)}$.

When sterile insects are released spatially uniformly, how can the shape and speed of the travelling wave of fertile insects be manipulated? To explore this question, we suppose first that both sterile and fertile insects are homogeneously distributed. The spatially uniform steady-state solutions to eqns (8.8) are given by the equilibria to the spatially independent version of eqns (8.8) (i.e., Barclay and Mackauer's model). Thus by analysing the intersections of the null clines, we can show that there always exists an equilibrium $E_0 = (0, (-A + \sqrt{A^2 + 8R})/4)$, and if $R \le R_c$, there additionally exist two other positive equilibria, $E_+ = (u^+, n^+)$ and $E_- = (u^-, n^-)$, where $u^- \le u^+$ and $n^+ \le n^-$ and R_c is the positive root of $108R^2 + 4(1 + A)\{A - 2(1 - A)^2\}R - A^2(1 - A)^2 = 0$ (see Fig. 8.4). E_0 represents the extinction of fertile insects, and E_+ represents the outbreak of fertile insects.

Linear stability analyses of eqns (8.8) about these three equilibria show that both E_0 and E_+ are locally stable to spatially homogeneous perturbations, while E_- is unstable. Thus Lewis and van den Driessche predicted

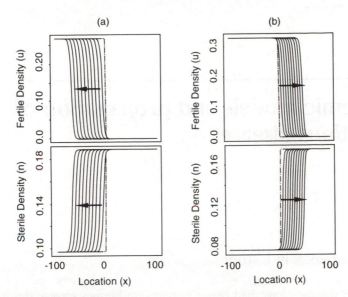

Fig. 8.5 Travelling wave solution of model (8.8), joining E_+ to E_0 for $A = \delta = 0$ (redrawn from Lewis and van den Driessche, 1993). Dashed line indicates initial distributions. (a) $R = R_0 + 0.005$, retreating wave of fertile extinction; (b) $R = R_0 - 0.005$, advancing wave of fertile invasion, where $R_0 = 0.0666$ ($< R_c = 0.0741$). (See text for details.)

that if R (or alternatively, the release rate of sterile insects) exceeds R_c, then the outbreak steady state is eliminated and the system approaches the extinction steady state E_0. On the other hand, when $R \leq R_c$, they numerically showed that there exist travelling wave solutions that join the outbreak steady state E_+ to the extinction steady state E_0. By analysing the speed of the travelling wave, they found that there exists another critical value R_0 which is smaller than R_c: when $R_0 \leq R \leq R_c$, the travelling wave retreats, leading to local extinction; conversely, when $R \leq R_0$, the travelling wave advances, resulting in an invasion (see Fig. 8.5). Interestingly, as Lewis and van den Driessche have noted, when $R_0 < R \leq R_c$, eqns (8.8) predict a travelling wave of extinction, while the model neglecting insect dispersal (i.e., spatially independent version of eqns (8.8)) predicts the antithesis—persistence of an outbreak.

9

Epidemic models and propagation of infectious diseases

9.1 Epidemics in history

Infectious diseases have had major impacts on human society throughout its history. Many medieval European cities are known to have declined in the end because of infectious diseases. The number of deaths caused by infectious diseases far exceeds that of all wars in the past. For instance, bubonic plague that swept over Europe in the 14th century took the lives of a quarter of the total population of 100 million. In the Aztec Empire of Central America, smallpox which was brought over by the Europeans reduced the population by half from about 3.5 million originally present in 1520. In this century, a strain of influenza known as Spanish flu broke out at the end of the First World War and caused the death of 20 million people around the world in 1919 (Anderson and May, 1991).

Diseases have greatly affected plants and animals as well. Dutch elm disease, caused by the fungus *Ceratocystis ulmi* and spread through its carrier, the European elm bark beetle, blighted extensive stands of elm trees throughout Europe. The fungus *Cryphonectria parasitica* that causes chestnut blight entered the United States from Asia at the beginning of this century. It spread on average almost 40 km per year and decimated the America chestnut as a dominant tree species in the eastern United States within 50 years.

As for animals, the rinderpest that entered Somalia in Africa in 1889 is most notorious. This is thought to have been introduced by Zebu cattle brought in by the Italian army; the virus rapidly spread among native ungulates such as buffalo and antelopes. The epidemic wave propagated from Somalia to the Cape in 10 years, as shown in Fig. 9.1. The speed of the spread was about 500 km per year (Dobson and May, 1986a). In southern Zambia, 90% of the buffalo population was wiped out in just two

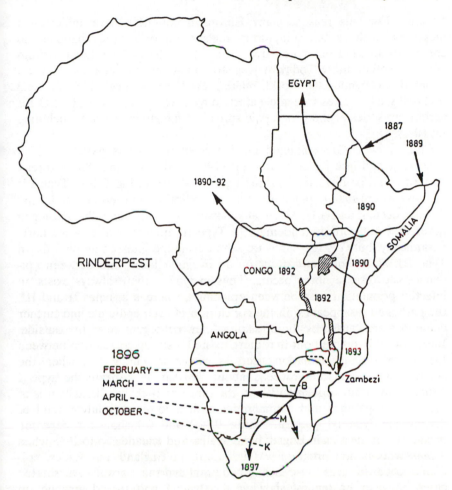

Fig. 9.1 Geographic spread of rinderpest in Africa, 1889–97 (after Mack, 1970).

years. One side effect was that the survival rate of wildebeest doubled. As this shows, aside from reducing the population of the afflicted species, infectious diseases often greatly affect other species that are directly or indirectly interrelated with that species (Dobson and May, 1986b).

While the agents of infectious diseases are mostly viruses, bacteria or protozoans, invasion by such pathogens does not always cause a breakout. Whether a certain disease will spread or not depends not only on the pathogen's contagiousness but also on the population size of susceptibles in the area and on their social structure. Let us examine this aspect using measles as an example.

Measles is known as an extremely contagious infectious disease affecting humans. For this reason, many European countries have undertaken measures, such as the compulsory registration of measles patients, to combat its spread from quite early on (e.g., from the 1880s in Great Britain and the Netherlands). Today the registration system for measles has been adopted throughout the world, making available detailed data on local outbreaks of measles (see the entertaining text of Cliff *et al.* (1981) for various episodes of the geographic spread of infectious diseases including measles).

Cliff *et al.* (1981) examined incidences of measles in various British cities reported in a paper by Bartlett (1957) and found that the outbreak pattern could be divided into three types as shown in Fig. 9.2(a). Type I is the case where major epidemics occur periodically; between epidemics the disease occurs regularly, albeit at a lower frequency, and the pathogen never dies out entirely (endemic). In Type II, epidemics also occur fairly periodically, but very few cases are seen during the intervening periods. In Type III, small epidemics occur on an irregular basis, and between epidemics the disease never occurs. Thus in type I, there always exists an infected population even between epidemics, whereas in types II and III, the pathogen disappears with the conclusion of each epidemic and further epidemics are caused by the invasion of new pathogens from the outside. Bartlett (1957) pointed out that there existed a strong correlation between the patterns and population size. As shown in Fig. 9.2(b), when the community's population exceeds about 200,000, it displays the type I pattern; when the population lies in the range of 10,000 to 200,000, it is of the type II pattern; and communities with a lower population tend to display the type III pattern. Figure 9.3(a) and (b) show the reported incidences of measles in England and Wales, and Iceland (up to 1968 when a mass vaccination program was initiated). In England and Wales, epidemics occur in recurrent cycles of two years and there are always isolated cases between epidemics, showing the type I pattern. Meanwhile, in Iceland epidemics occur every four years but between epidemics there are often no cases, putting it under the type II pattern. As seen here, many infectious diseases among humans exhibit recurrent cycles in incidence, with the period varying with the population density. Thus, in terms of the average period, whooping cough (pertussis) has had recurrent cycles of 3 years in England and Wales, and 3.2 years in Baltimore, while chicken pox had an epidemic cycle of 2.5 years in New York and 3.0 years in Glasgow, and measles, 3.3 years in Glasgow and 3.4 years in Baltimore. Shaffer and Kot (1985) have pointed out, however, that the incidence patterns are not always regular but are somewhat chaotic (see also Sugihara *et al.*, 1990).

Anderson and May (1986), who have contributed greatly to the study of mathematical models on the ecology of epidemics, have made a comparative study of infectious diseases and noticed that when a major epidemic occurs, it follows the following three stages:

(a)

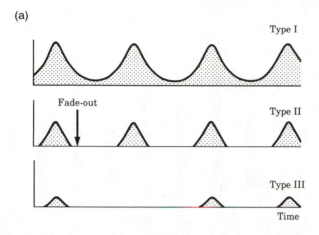

(b)

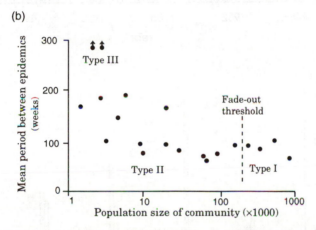

Fig. 9.2 (a) Three types of measles epidemic profile: type I has an endemic pattern with periodic eruptions; type II has a regular pattern of epidemics with fade-outs; type III has an irregular pattern of epidemics with occasional fade-outs (Cliff *et al.*, 1981). (b) Impact of population size on the spacing of measles epidemics for 19 English towns (from Cliff *et al.*, 1981, based on data compiled by Bartlett, 1957).

(1) initial establishment: a pathogen which has invaded a non-infected region continues to increase in number, thus establishing a foothold for later epidemics;
(2) persistence: the locally distributed pathogen persists within the host population in the long term;
(3) spatial spread: the pathogen spreads to other non-infected regions, thus expanding its spatial spread.

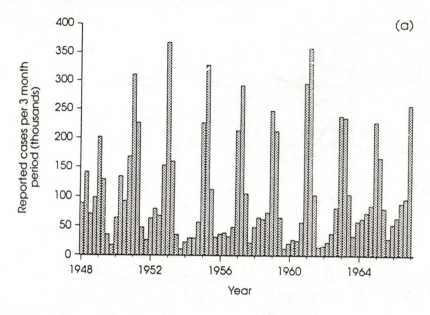

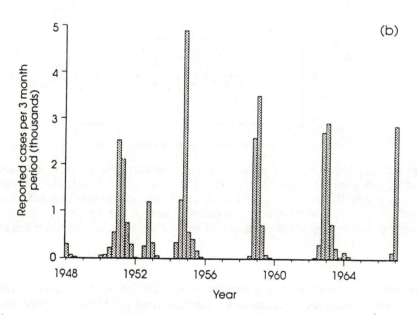

Fig. 9.3 Recurrent epidemics of measles observed during 1948–66 (after Anderson and May, 1986). (a) Reported cases in England and Wales. Measles virus persists endemically, showing recurrent two-year cycles in incidence. (b) Reported cases in Iceland, Epidemics occurs at roughly four-year intervals and between epidemics the virus vanishes from the island.

In the following sections, we examine the conditions under which stages (1) and (2) occur mostly in the case of measles, based on the analyses by Kermack and McKendrick (1927) and Anderson and May (1979, 1986, 1991). The spatial spreading of disease, which is stage (3), will be discussed in section 9.4 for the bubonic plague, and in the next chapter for the case of rabies.

9.2 Kermack–McKendrick model

We first introduce the classical model of Kermack and McKendrick (1927), which is simple but captures the essence of the issue of initial establishment.

Let us consider a situation in which a few infectives (I_0) arrive at an area with population size S_0 which has never been previously invaded by the pathogen. The infectives come into contact with and infect susceptible individuals, and meanwhile a certain proportion either die from the disease or recover and develop immunity. If we denote the number of the susceptibles and infectives at time t by $S(t)$ and $I(t)$, respectively, they vary according to the following equations:

$$\frac{\mathrm{d}S}{\mathrm{d}t} = -\beta SI, \tag{9.1a}$$

$$\frac{\mathrm{d}I}{\mathrm{d}t} = \beta SI - \gamma I. \tag{9.1b}$$

Here, βSI is the rate by which the infectives increase, and this is assumed to be proportional to the number of contacts made between susceptibles and infectives (i.e. susceptibles and infectives are homogeneously mixed). The proportionality constant β is called the transmission coefficient, while γI represents the rate at which infectives disappear, either by host mortality or recovery from disease.

Since we are considering the case where initially I_0 infectives invade a susceptible population with size S_0, the initial values of eqns. (9.1) are given by

$$\begin{aligned} S(0) &= S_0, \\ I(0) &= I_0, \end{aligned} \tag{9.2}$$

where I_0 is sufficiently smaller than S_0.

The number of infectives must increase from its initial value for the disease to become established in the beginning. For the infective population to be increasing at the outset, the right-hand side of eqn (9.1b) must be positive, or $(\beta S_0 - \gamma)I_0 > 0$. Thus, the condition for initial establishment is given by

$$S_0 > \frac{\gamma}{\beta} \equiv S_t. \tag{9.3}$$

If, conversely, the initial susceptible population S_0 is lower than γ/β, the invading disease will not be able to spread. $S_t = \gamma/\beta$, defined by the right-hand side, is called the 'threshold population size' of susceptibles. The higher the transmission coefficient β, and the lower the mortality (or recovery) rate γ, the smaller is the threshold population size S_t. From eqn (9.3), we see that to stop an epidemic from occurring, we need only to make S_t greater than S_0, by either lowering the susceptible population S_0 to below S_t or lowering β by, say, isolating the infectives. In the case of human diseases, the former method is frequently employed, where the susceptible population S_0 is lowered by vaccination or other means. Here, note that to control the epidemic it is unnecessary to vaccinate the entire population but only needed to carry out vaccination until the susceptible population falls below S_t. In fact, due to effective vaccination campaigns, smallpox has been virtually wiped out, while cases of diphtheria and polio have also become rare in Western nations today. On the other hand, the pathogenic germs that cause measles or whooping cough, which are transmitted by droplet infection and thus highly contagious, have very low S_t values and so have not been successfully controlled so far. In order to eradicate measles, in particular, it is said that over 90–95% of the population must receive vaccination (Anderson and May, 1991). Thus, in large cities, where there is a continual inflow and outflow of people, a vaccination programme cannot be expected to be very successful.

We now examine intuitively the ecological significance of eqn (9.3), which is rewritten as

$$\Gamma \equiv \frac{\beta S_0}{\gamma} > 1, \tag{9.4}$$

where $1/\gamma$ is the average duration that an individual remains infectious (i.e., life expectancy of an infective), while βS_0 is the number of secondary cases generated by an infective per unit time. Their product, $\Gamma = \beta S_0/\gamma$, can therefore be considered as the number of secondary cases of infection generated by an infectious individual. In this sense, Γ is called the disease's 'basic reproductive rate'. Equation (9.4) or its equivalent, eqn (9.3), expresses the obvious fact that if the number of secondary cases that a single infective generates is greater than unity, the disease will succeed in initially establishing itself. For measles, it has been found that $\Gamma = 13-18$ in Western societies, that is, a single primary case causes 13–18 secondary cases.

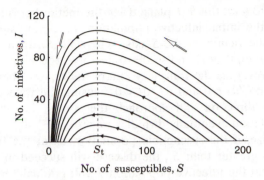

Fig. 9.4 Phase plane diagram of the Kermack–McKendrick model. S and I are population sizes of susceptibles and infectives, respectively. Infectives will increase only when the population size of susceptibles is larger than the threshold S_t as indicated by the dashed line. $S_t = \gamma/\beta = 50$.

So far, we have focused only on whether a disease will increase in the initial stages. What kind of changes will a disease undergo later, after it has successfully established itself? This can be seen by examining the

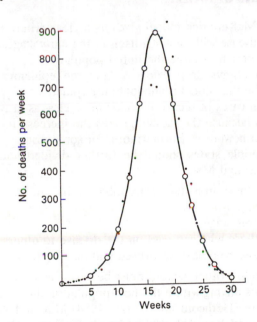

Fig. 9.5 Deaths from plague in the island of Bombay over the period 17 December 1905 to 21 July 1906 (solid dots). Solid line and open circles represent the solution of the Kermack and McKendrick model (after Kermack and McKendrick, 1927).

solution trajectory on the $S-I$ plane using the method of isoclines (see Fig. 9.4). Because the initial infective population, I_0, is very small, the trajectory starts in the vicinity of the S-axis. When $S > S_t$, then $dS/dt < 0$ and $dI/dt > 0$ from eqns (9.1) and so the trajectory proceeds in the upper left direction. However, as the trajectory enters the region where S is smaller than S_t, then $dS/dt < 0$ and $dI/dt < 0$, causing the trajectory to shift to the lower left direction and approach the S-axis; in other words, the infective population eventually dwindles to zero.

To summarize, if the susceptible population S_0 at the time of disease introduction is greater than S_t, the disease will succeed in initially establishing itself and the infective population I will gradually increase; subsequently, a peak population will be reached, after which the number of infectives will dwindle and the epidemic will eventually die down. As an example where S_0 was greater than S_t, Kermack and McKendrick cited the pest epidemic in Bombay during 1905–06 and showed that the change in the epidemic's intensity over time could be described well by their model (see Fig. 9.5).

9.3 Epidemics of measles

The Kermack–McKendrick model gives us a clear criterion for determining whether a disease will establish itself in the early stages. In this model, however, no matter how large the initial population S_0 is, the susceptible population will always drop below S_t and the epidemic will terminate eventually. While this model is adequate for epidemics of short duration, it does not explain cases of recurrent epidemic cycles, as in types I and II of Fig. 9.2. This is because the model ignores the increase in the susceptible population from newborns. Furthermore, for some disease types the infective and susceptible states should be further divided into the following stages (Anderson and May, 1982):

(1) susceptible: those in a state susceptible to disease infection;
(2) infected: those who are infected but latent; the contagious power is low during this stage;
(3) infectious: those who can pass on the disease to others;
(4) immune: those who have recovered and developed immunity.

In real situations, other factors must be taken into consideration, such as heterogeneous mixing within the host population due to the presence of a social structure (Hethcote and Yorke, 1984; May and Anderson, 1984, 1987; Hethcote and Van Ark, 1987; Castillo-Chavez, 1989), migrations in and out of the population (Pyle, 1969, 1982; Cliff *et al.*, 1981; see also sections 9.4 and 10.3) and vertical transmission from parent to offspring, and so the model must be adjusted accordingly for the situation at hand.

Here, with the example of measles in mind, we investigate the course that an epidemic will take in the long run, by employing an extended Kermack–McKendrick model as given by the following equations:

$$\frac{dS}{dt} = rN - bS - \beta SI,$$

$$\frac{dH}{dt} = \beta SI - (b + b_H + \sigma)H,$$

$$\frac{dI}{dt} = \sigma H - (b + b_I + \gamma)I,$$

$$\frac{dR}{dt} = \gamma I - bR,$$

(9.5)

where S, H, I and R are the number of susceptible, infected, infectious and immune individuals, respectively. $N = S + H + I + R$ is the total host population, r the birth rate, β the transmission coefficient, σ the rate at which an infected person passes from the latent state to the infectious state ($1/\sigma$: average latent period), γ the recovery rate ($1/\gamma$: average infectious period), b the death rate of susceptibles ($1/b$: the life expectancy of a healthy individual), b_H the increase in death rate caused by the disease at the infected stage, and b_I the increase in death rate at the infectious stage. Figure 9.6 shows a flow diagram for the four stages

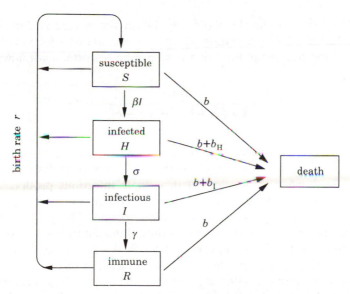

Fig. 9.6 Flow diagram of measles infection. Population is divided into four stages: susceptible S, infected H, infectious I and immune R.

described above. In other words, susceptibles become infected at the rate βI, who become infectious at the rate σ, and then recover and become immune at the rate γ.

For measles, the parameters have been estimated as follows: $1/\sigma$ (average latent period) = 6–9 days; $1/\gamma$ (average infectious period) = 6–7 days. In Western industrialized societies, incidences of death caused by measles are extremely rare, so that values of b_H and b_I are nearly equal to zero. The transmission coefficient β depends not only on the pathogen's contagious power but also on the frequency with which infectives mix with susceptibles; thus two population groups of identical size but with different life styles or habits will have different β values. In many cases, β is estimated indirectly from the critical threshold population size S_t or the disease's basic reproductive rate Γ.

We now take the case of eqns (9.5) and, as we did for the original Kermack–McKendrick equation, see whether the disease will successfully become established initially when a small number of infectives (either infected or infectious) invade a susceptible population of $S_0(=N_0)$. Thus using the linear stability analysis for eqns (9.5) in the vicinity of $(S, H, I, R) = (S_0, 0, 0, 0)$, we have the following criterion (see section 9.7):

$$S_0 > \frac{(\sigma + b + b_H)(\gamma + b + b_I)}{\beta\sigma} \equiv S_t. \tag{9.6}$$

S_t, defined thus, is the threshold population size necessary for establishment, which the host population S_0 must exceed in order for the disease to spread in the beginning. Equation (9.6) can be rewritten as follows:

$$\Gamma \equiv \frac{\beta\sigma S_0}{(\gamma + b + b_I)(\sigma + b + b_H)} > 1, \tag{9.7}$$

where Γ denotes the basic reproductive rate of the disease. Thus, $1/(\gamma + b + b_I)$ is the average duration that the infectious stage lasts after onset, βS_0 is the number of infecteds that a single infectious produces in unit time, and $\sigma/(\sigma + b + b_H)$ is the rate at which an infected becomes infectious without dying during the latent period. The product of these three quantities, Γ, represents among those infected by a single infectious, the number of individuals who pass through the latent period to become an infectious. If this is greater than unity, then the disease will increase. For infectious diseases affecting humans, $1/(\sigma + b + b_H)$ (average latent period) and $1/(\gamma + b + b_I)$ (average infectious period) are often from several days to several weeks. Meanwhile, the life expectancy $1/b$ of healthy individuals is around 70 years, so we can say that $\sigma \gg b$, $\gamma \gg b$. Thus, eqns (9.6) and (9.7) can be approximated as

$$S_0 > S_t \equiv \frac{\gamma}{\beta}, \qquad \Gamma \equiv \frac{\beta S_0}{\gamma} > 1.$$

These expressions are respectively identical to eqns (9.3) and (9.4), which were obtained from the Kermack–McKendrick model, confirming that the Kermack–McKendrick model is a good approximation with regard to initial establishment.

When the condition for initial establishment, eqn (9.6), is satisfied, what kind of behaviour will the solution of eqns (9.5) display with respect to time? Here we focus on the special case where the death rates caused by the disease are zero, i.e. $b_H = b_I = 0$, and the population size remains fairly stable so that $dN/dt = 0$, i.e. $r = b$ holds. Although the assumption of a fixed population size (i.e., $r = b$, $b_H = b_I = 0$) does not apply to areas undergoing rapid population growth, it is nevertheless often adopted to simplify analysis when studying infectious disease models for human populations. In the next chapter, however, we will look at infectious diseases among animals where this assumption does not hold.

Using numerical computation, we find that S, H, I and R all approach a positive equilibrium point while oscillating (note that a positive equilibrium point for eqns (9.5) always exists when eqn (9.6) is satisfied). A typical trajectory is depicted in Fig. 9.7. This figure suggests that the recurrent cycles observed in measles correspond to this oscillatory state before

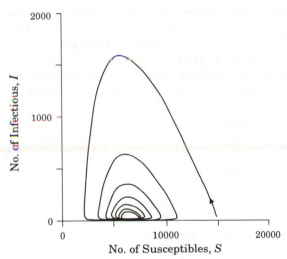

Fig. 9.7 A solution of the *SHIR* model, eqns (9.5), projected on the (S, I) plane. $1/\sigma = 7$ days, $1/\gamma = 7$ days, $1/b = 70$ years, $\beta = 0.008$.

Table 9.1 Transition events and corresponding rates for stochastic model of measles virus transmission

	Type of transition event				Transition rate	Event
(1)	$S \to S + 1$,	$H \to H$,	$I \to I$,	$R \to R$	rN	birth
(2)	$S \to S - 1$,	$H \to H$,	$I \to I$,	$R \to R$	bS	death
(3)	$H \to H - 1$,	$S \to S$,	$I \to I$,	$R \to R$	bH	death
(4)	$I \to I - 1$,	$S \to S$,	$H \to H$,	$R \to R$	bI	death
(5)	$R \to R - 1$,	$S \to S$,	$H \to H$,	$I \to I$	bR	death
(6)	$S \to S - 1$,	$H \to H + 1$,	$I \to I$,	$R \to R$	βSI	infection
(7)	$H \to H - 1$,	$I \to I + 1$,	$S \to S$,	$R \to R$	σH	becoming infectious
(8)	$I \to I + 1$,	$S \to S$,	$H \to H$,	$R \to R$	Λ	immigration
(9)	$I \to I - 1$,	$R \to R + 1$,	$H \to H$,	$S \to S$	γI	recovery

Source: Anderson and May (1986)

converging to a steady state. Upon closer inspection of Fig. 9.7, however, we notice that during each oscillatory cycle I goes through a phase when it is very nearly zero. When I is very low, the chances are high that the pathogen will become extinct due to stochastic effects. Taking notice of this fact, Anderson and May (1986) rewrote eqns (9.5) as an equivalent stochastic equation. In other words, assuming that S, H, I and R follow the transition rules given in Table 9.1, they did a computer simulation using the Monte Carlo method. Simulations were carried out for the measles epidemics in England and Iceland, which were shown in Fig. 9.3(a) and (b). They used $S_0 = N = 250,000$, $S_t = 15,000$ and $r = 3571$ for England, and $S_0 = N = 100,000$, $S_t = 6,500$ and $r = 1428$ for Iceland, with the other parameters chosen in common as $1/\sigma = 7$ days, $1/\gamma = 7$ days, and $1/b = 70$ years. In addition, they assumed that the number of infectious individuals entering from abroad was $\Lambda = 7$ per year. The simulated results, shown in Fig. 9.8(a) and (b), show a close match with the observed data; England displays the type I pattern, with epidemic and endemic states recurring in about two-year cycles, while Iceland, with a much lower population, shows type II characteristics, with epidemics occurring in roughly four-year cycles with intermissions when there are no cases.

Notice that even though Iceland has a low population, it still satisfies eqn (9.6), the condition for initial establishment. According to the deterministic model of eqns (9.5), therefore, the number of infections goes through oscillatory swings while approaching a positive equilibrium point. However, the stochastic model allows for the possible extinction of the disease by stochastic effects when the epidemic is at a low ebb in the cycle. Even when the disease dies down, when the susceptible population recovers its numbers by the addition of newborns and exceeds S_t, the invasion of an infective from without will trigger another epidemic outbreak. The type I pattern, in which recurrent cycles of epidemics occur without

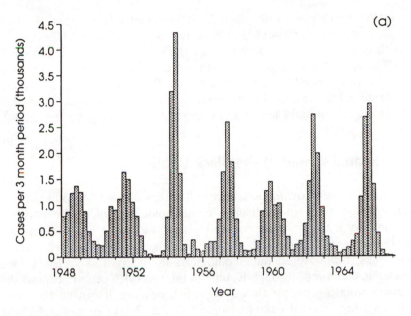

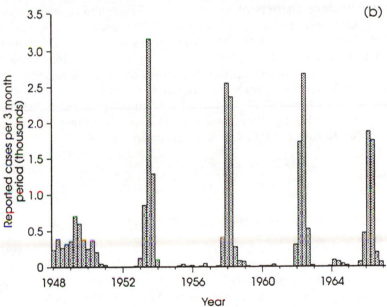

Fig. 9.8 Predicted cases of measles calculated from the stochastic model (after Anderson and May, 1986). Parameter values: (a) $N = 250,000$, $S_t = 15,000$, $r = 3571$; (b) $N = 100,000$, $S_t = 6500$, $r = 1428$. Other parameters are set in common as $1/\sigma = 7$ days, $1/\gamma = 7$ days, $1/b = 70$ years and $\Lambda = 7$ per year. Results of simulation for (a) and (b) show a close match with the observed data in England and Iceland as given by Fig. 9.3(a) and (b), respectively.

extinction, takes place only when S_0 is sufficiently larger than S_t. In the measles case, S_t is considered to be between 200,000 and 300,000.

Although we have introduced mostly the models presented by Anderson and May, there are many other models that have been mathematically analysed and applied to infectious diseases. Readers who are interested in the details of these models are encouraged to read the reviews by Bailey (1975), Anderson and May (1991), Capasso (1993) and Hethcote (1994).

9.4 Spatial spread of the Black Death

As we have seen so far, pathogens with higher basic reproductive rates (transmission rates in particular) cause epidemics more frequently. In addition to measles, influenza and plague, which are also respiratory diseases transmitted through air and thus have extremely high transmission rates, have caused major epidemics in the past. The Black Death pandemic that swept through Europe in the mid-14th century caused sheer terror among the people (it was so called because it caused the skin to blacken from internal haemorrhaging). It first broke out in Italy, where Venice lost three-quarters of its population. Subsequently, major cities in France, Germany and England underwent a similar fate. In the short period of seven years, the number of deaths reached a quarter of the total population of Europe. Figure 9.9 shows the wave front of the plague from 1347 to 1350 as it moved north from Italy and reached Scandinavia. It shows that the spreading wave travelled at 320–650 km per year.

Nobel (1974) theoretically explained the spread pattern of Fig. 9.9 by using the following equation, which is obtained by adding the diffusion term to the Kermack–McKendrick model of eqns (9.1):

$$\frac{\partial S}{\partial t} = D \frac{\partial^2 S}{\partial x^2} - \beta S I,$$

$$\frac{\partial I}{\partial t} = D \frac{\partial^2 I}{\partial x^2} + \beta S I - \gamma I.$$

$$(9.8)$$

$S(x,t)$ and $I(x,t)$ are the densities of susceptibles and infectives, respectively, and D is the diffusion coefficient to account for the migration and spread of people; other coefficients are the same as in eqns (9.1). Note that these equations are analogous to the Lotka–Volterra equations with diffusion terms, (8.1), if predators and prey correspond to infectives and susceptibles, respectively, and susceptibles give no birth.

Based on the speed at which news and rumours travelled at the time, the diffusion coefficient is estimated at $D = 2.6 \times 10^4$ km^2/year. The mortality rate is $\gamma = 15$/year. Although the transmission rate β is the most difficult to estimate, we may set a value at $\beta = 1$ km^2/year considering the

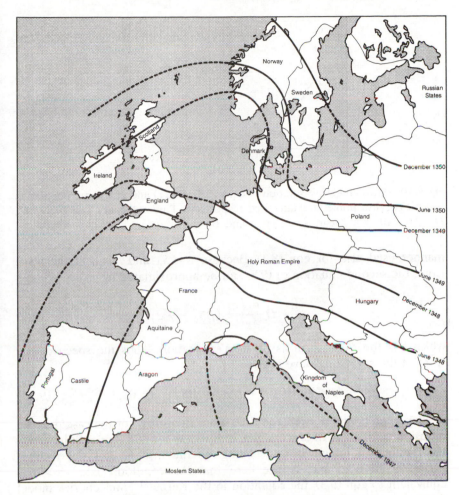

Fig. 9.9 Spread of Black Death in Europe during 1347–50 (after Langer, 1964).

transmission route from rats to humans via the agent of fleas. The population of Europe in 1347 immediately before the plague broke out was 85 million, which translates to a density of 19.5 per km^2. Thus, as the initial condition, the density of the susceptible population was uniformly set to be $S(x, 0) = S_0 = 19.5/\text{km}^2$ throughout the entire area. Then assuming that the pathogen was introduced near the origin, eqns (9.8) were numerically solved. As shown in Fig. 9.10, the results show that $I(x, t)$ and $S(x, t)$ both propagate at the same constant speed, the former as a single peak and latter as a sigmodial curve. Since S_0 less S and I is the density of those who died, it is not hard to visualize the heap of dead bodies left behind after the peak of I had passed through.

The speed of the propagating wave is now determined by the intuitive

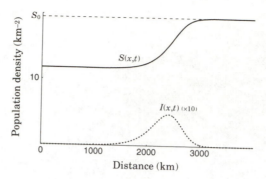

Fig. 9.10 Travelling wave solution of eqns (9.8). $S(x, t)$ and $I(x, t)$ are population densities of susceptibles and infectives, respectively. Parameters are $S_0 = 19.5/\text{km}^2$, $D = 2.6 \times 10^4 \text{ km}^2/\text{year}$, $\gamma = 15/\text{year}$ and $\beta = 1 \text{ km}^2/\text{year}$.

manner used also for eqn (8.3). Because $S \approx S_0$ and $I \approx 0$ at the wave front, the second equation of (9.8) can be approximated by

$$\frac{\partial I}{\partial t} = D \frac{\partial^2 I}{\partial x^2} + (\beta S_0 - \gamma)I, \tag{9.9}$$

which has the same form as Skellam's equation. Thus the speed of the front of the spreading wave is given by

$$c = 2\sqrt{D(\beta S_0 - \gamma)} \text{ for } \beta S_0 - \gamma > 0. \tag{9.10}$$

(Kallen *et al.* (1985) derived eqn (9.9) more rigorously for the case that the diffusion coefficient of susceptibles is zero.) When $\beta S_0 - \gamma < 0$, however, the number of infectives is declining, so the disease dies out without becoming an epidemic. Notice here that the condition $\beta S_0 - \gamma > 0$ is equivalent to eqn (9.3), the condition in the Kermack–McKendrick model for a disease's initial establishment. Thus, we see that the judgement criterion for whether a certain disease will spread or not still rests on whether the basic reproductive rate Γ $(= \beta S_0/\gamma)$ is greater or smaller than unity, and not on the dispersal ability. Mollison and Kuulasmaa (1985) noted, however, that if dispersal occurs stochastically on a two-dimensional lattice, the invasion criterion is somewhat altered (see also Tainaka, 1988; Sato *et al.*, 1994; Kawasaki *et al.*, 1996).

When the above estimated values for the parameters are substituted into eqn (9.10), we find that indeed $\beta S_0 - \gamma > 0$ is satisfied to give a speed of $c = 684 \text{ km/year}$. This result agrees well with the observed spread rate of 320–650 km/year.

In 1665, a plague epidemic broke out once again in London, and the destruction wrought then is vividly recounted in novels and poems from the time as well as in an English nursery rhyme. The latest plague pandemic occurred in 1850 in China, with total casualties of about 13 million; according to the 1959 WHO report, it was officially finished only in 1959. Although there is a widely held belief that plague is no longer a problem, Murray (1989) reports on the existence of residual foci of plague among mice and voles in the western United States, particularly New Mexico, and in Russia, and warns us of the serious potential of a plague epidemic in the eventuality that these animals come into contact with large, concentrated urban populations. In fact, the outbreak which occurred near Bombay in 1994 is still fresh in our memory.

Equations (9.8) and (9.10) were also applied to the spread of rabid fox (Fig. 10.1) by Kallen *et al.* (1985) and rinderpest (Fig. 9.1) by Dobson and May (1986a). They estimated the rates of spread and also the required width of a 'fire break' (where susceptibles are removed) which prevents the spread of the epidemic (see section 10.4).

9.5 The effects of epidemics on population demography

Focusing on the fact that many human diseases are maintained as an endemic infection only in communities with a high population density, May (1985) discussed the historical relationship between disease and human population.

Prior to about 10,000 years ago, human communities everywhere were hunter-gatherer societies, and the population density was very low. Thus, epidemics of smallpox, measles, cholera, etc. could not have existed. These and other epidemic diseases gradually took hold with the establishment of agriculture some 10,000 years ago. Indeed, if we look at the earth's population over the past 10,000 years, we see that the first 5000 years saw the population rise from five million to 100 million, a 20-fold increase, whereas from 5000 years ago to 400 years ago the population grew from 100 million to 500 million, showing a much smaller (5-fold) increase. Previously, this decline in growth rate has been thought of as a levelling-off caused by the overconcentration of population, but an important factor must have been the increased incidences of infectious diseases as cities came into existence whose populations exceeded the epidemic threshold population.

An example that strikingly illustrates the relation between disease and human populations can be seen in the European invasion of the New World. The diseases introduced by Europeans served in effect as biological weapons, and the native populations of the American continent were reduced to 5% of their original level before any hostile engagements ever took place. The Europeans in turn were little affected by the native

American diseases with the possible exception of syphilis. Why did such a one-sided invasion of disease take place? Why did the New World's native population lack resistance, which the Europeans possessed? May provides the following explanation. The native populations of America had very low densities even before the European arrival, densities that were too low to allow the initial establishment of infectious diseases. Having little resistance against such diseases, they were decimated when invaded by diseases 'tempered' under harsh European conditions. Syphilis is one of very few diseases that have a long infectious period (i.e., γ is small) and so is able to remain infectious even in human populations with low density.

9.6 Evolution of myxoma virus in the European rabbit

So far in our discussion, we have assumed that the characteristics of the pathogen and host remain constant. There are many known examples, however, where the host–pathogen relationship has moved toward a higher adapted state. It has been known for some time, for instance, that when a

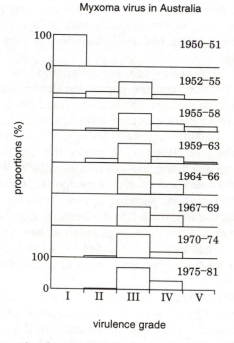

Fig. 9.11 Proportions of various grades of myxoma virus found in wild populations of rabbits in Australia at different times from 1951 to 1981 (after May and Anderson, 1983).

newly introduced species is afflicted by a local disease, it is often devastated to near extinction; yet after the passage of some time, the pathogen loses some of its virulence, or the host acquires a higher resistance, and the disease becomes relatively harmless to its host. A case in point is provided by the co-evolution of the European rabbit and myxoma virus in Australia, which has been well documented based on abundant field data and extensive laboratory experimentation.

The European rabbit was introduced into Australia in 1859 as a game animal. In 20 years time, however, it had bred and multiplied to the point where it had become a major pest for the grazing industry. Thus, in an attempt to curtail its fecundity, in 1950 the Australian government introduced into the rabbit population the myxoma virus, which was endemic to the tropical rabbits of South America, against which it is relatively harmless. Against the Australian rabbits, however, it showed an extremely high virulence, killing 99.8% of the infected rabbits. Yet within two to three years of introduction, the virus gradually lost its virulence and the death rate of the rabbits fell. From laboratory experiments, it was found that this was because after the initial introduction of a highly virulent strain (grade I) there rapidly appeared a series of successively less virulent strains (grades II, III, IV and V in descending order of virulence), and that 30 years after introduction, the strains with intermediate grades of virulence (grades III and IV) had become the dominant ones. (See Fig. 9.11.)

May and Anderson (1983) have claimed that a pathogen's evolution moves in a direction so as to increase its basic reproductive rate as defined by:

$$\Gamma = \frac{\beta S_0}{\gamma + b + b_{\mathrm{I}}}. \tag{9.11}$$

Equation (9.11) represents the limit state for eqn (9.7) when the latent period is zero (i.e., when $\sigma \to \infty$). Thus, if only the pathogen's virulence b_{I} is allowed to vary, the basic reproductive rate Γ is greater for smaller b_{I}. Hence, as the argument of May and Anderson goes, the disease will tend to a state of harmlessness. This does seem to provide an explanation for the old notion that diseases eventually become harmless. In the case of myxoma virus, however, it has been established by laboratory experiments that, as the virulence b_{I} changed, so did the transmission rate β and recovery rate γ. Thus, virulence b_{I} has a positive correlation with transmission rate β and a highly negative correlation with recovery rate γ, as shown in Fig. 9.12(a). May and Anderson substituted the γ–b_{I} relationship of Fig. 9.12(a) into eqn (9.11) (where, due to lack of sufficient data, β was taken as a constant) to see how the basic reproductive rate Γ is affected by b_{I}; the result is shown in Fig. 9.12(b). It shows that Γ is maximized for a

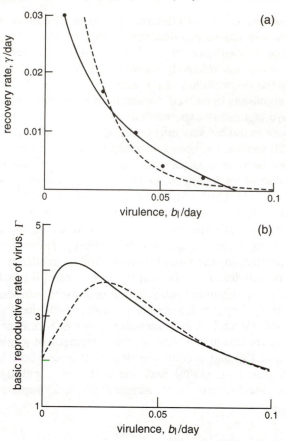

Fig. 9.12 (a) Relationship between recovery rate γ and virulence b_I for various strains of myxoma virus in wild population of rabbits in Australia. Solid dots are observed data. Solid and dashed lines are regression curves for $\gamma(b_I) = c + d \ln b_I$ and $\gamma(b_I) = c \exp(-db_I)$, respectively. (b) Relationship between basic reproductive rate Γ and virulence b_I for functional relations between γ and b_I displayed in (a). $\beta = 0.011/\text{day}$. (May and Anderson, 1983.)

pathogen of intermediate virulence; too large a b_I kills off the rabbits too quickly, reducing the chances of transmission; too small a b_I leads to rabbits recovering too quickly, again lowering the transmission efficiency. This analysis explains why, in Fig. 9.11, myxoma virus strains with intermediate virulence remained.

Levin and Pimentel (1981), on the other hand, examined the case where two strains of myxoma virus with different virulence levels were present; they analysed the dynamics of a host population of rabbits when a host already infected by the weaker strain becomes infected with a more virulent one. Their results showed two possible outcomes: the virus with higher basic reproductive rate remains, as claimed by May and Anderson; or, in cases where the two strains do not differ greatly in virulence level, the two strains coexist.

While the above models are concerned with the population dynamics of a host community, recent years have seen the arrival of various studies and models based on new approaches. One such study attempts to determine the pathogen's optimum reproductive schedule by viewing the pathogen's behaviour within the host's body as its life history strategy (Sasaki and Iwasa, 1990). Another model takes into consideration the pathogen's density in the environment (Kakehashi and Yoshinaga, 1992). In addition, several theories have been put forward concerning the effects that diseases have had on a host species' behaviour, for example, theories suggesting that diseases contributed to the evolution of sexual reproduction (Hamilton, 1980; Hamilton *et al.*, 1990) or of sexual dimorphism (Hamilton and Zuk, 1982). No doubt, the patterns of spatial spreading shown by an invading species, the main subject of this book, are also affected by the evolution of diseases, and investigations in this direction should become an important research topic in the future.

9.7 Appendix: threshold population for epidemic occurrence—derivation of eqn (9.6)

We determine the condition for which a small number of infectious that invade an initial state $(S_0, 0, 0, 0)$ will not go extinct in the beginning. Thus by putting $S = S_0 + s$, $H = h$, $I = u$ and $R = v$, and taking the linear approximation of eqns (9.5) in the vicinity of $(S_0, 0, 0, 0)$, we obtain equations for h and u in a closed form:

$$\frac{dh}{dt} = -(b + b_H + \sigma)h + \beta S_0 u,$$

$$\frac{du}{dt} = \sigma h - (b + b_I + \gamma)u.$$

The characteristic equation of the coefficient matrix for the right-hand side is given by

$$\lambda^2 + (2b + b_H + b_I + \sigma + \gamma)\lambda + (b + b_H + \sigma)(b + b_I + \gamma) - \sigma\beta S_0 = 0.$$

For neither h nor u to become extinct, the real part of at least one of the eigenvalues of the characteristic equation should be positive. Thus we have the following equation:

$$(b + b_H + \sigma)(b + b_I + \gamma) < \sigma\beta S_0,$$

from which eqn (9.6) follows directly.

10

Invasion of rabies in Europe

10.1   Recent epidemics in Europe

Rabies is known for its extremely high death rate and for the violent swings in the host population as the epidemic progresses. In this chapter, we examine the invasion and spatial spread of rabies.

A person who is infected with rabies dies in a short period after experiencing convulsions and other violent nervous symptoms, and for this reason, it has been feared since early times. Although in industrialized nations the incidence of rabies infections among humans is rare today, with an annual average of 4.3 cases in Europe, over 15,000 cases are reported worldwide every year (Anderson *et al.*, 1981). In Europe, which has high host populations of red fox, badgers and roe-deer, rabies has spread from wildlife to domestic animals, and considerable effort is expended on its control.

The most recent rabies epidemic began in Poland in 1939. Its range has spread at the rate of 30–60 km per year, with the advancing front currently in France (see Fig. 10.1). Since the United Kingdom has an even higher density of host populations of foxes than the European continent, once rabies crosses the Dover Straits, there is the potential for its becoming a major epidemic there. Hence, active research is being undertaken in the UK to predict the propagation route of rabies if it succeeds in entering, and on measures to control its spread (Macdonald, 1980; Anderson *et al.*, 1981; Kallen *et al.*, 1985; Bacon, 1985; Murray *et al.*, 1986; Murray and Seward, 1992).

In particular, Anderson *et al.* (1981) gathered extensive data on the ecology of the red fox (*Vulpes vulpes*), a major host species of rabies, and using this as a basis, attempted to explain theoretically the dynamics of the epidemics that had taken place. We introduce this study in the next section, then present the analyses on the spatial spread of rabies by Murray *et al.* (1986) and Yachi *et al.* (1989), and finally discuss the possibility of controlling rabies by constructing a 'firebreak' as proposed by Murray *et al.* (1986).

Fig. 10.1 Spread of rabies in Europe. Shaded region indicates western frontiers of European rabies in 1983. Figures with arrows are the dates when the epizootic crossed national borders (after Bacon, 1985).

10.2 Red fox and rabies

We first look at the ecology of red fox. Mature red foxes normally live in pairs, with each pair holding a territory of 2.5–16 km². In temperate zones, a vixen gives birth to a litter of 1–10 cubs (average of 4.7) once each spring. The offspring leave their parents the following autumn, although it is estimated that only about two offspring from a female parent succeed in establishing their own territories. The population dynamics of red fox can be thought to follow the following logistic equation:

$$\frac{dS}{dt} = (a - b - \mu S)S, \tag{10.1}$$

where S is the population density of red fox, a is the number of offspring (female) (about one per year) that a vixen can raise when fox density is low, and b is the death rate ($1/b$ is the life expectancy, which is about two years); μ is the intraspecific competition coefficient, representing the effect of deaths from territorial competition, etc. From eqn (10.1), the carrying capacity K is given by

$$K = \frac{a - b}{\mu}. \tag{10.2}$$

K is about 0.1–4 foxes per km^2 in Europe, although it can be considerably higher in suburban areas of the UK (Macdonald, 1980).

Consider now a situation where, into this healthy population of foxes, a few infectious ('rabid') foxes are introduced. As in the previous chapter, we divide the fox population into susceptible, latent (i.e., infected but not infectious) and rabid foxes, with their population densities denoted by S, H and I, respectively. Because the death rate of infected foxes is extremely high, the number of recovered, immune foxes is considered to be nil here (but see Murray and Seward (1992) for the effect of immune foxes). Note that, unlike the previous chapter, here S, H, and I represent the number of foxes per unit area. The dynamics of the transmission process of rabies can be represented, in a manner similar to model (9.5), by the following equations (see Fig. 10.2):

$$\frac{dS}{dt} = (a - b)S - \mu SN - \beta SI,$$

$$\frac{dH}{dt} = \beta SI - (b + \sigma + \mu N)H, \tag{10.3}$$

$$\frac{dI}{dt} = \sigma H - (b + b_1 + \mu N)I.$$

Here, $N = S + H + I$ (total population density), β is the transmission rate ($1/\beta$ is the average time period before a rabid fox encounters another fox: 4–5 days for a density of 1 fox per km^2), σ is the rate of clinical onset of disease ($1/\sigma$ is the average latent period: 28–30 days), and b_1 is the death rate of rabid foxes ($1/b_1$ is their life expectancy: 5 days). Since the life expectancies of latent and rabid foxes are both extremely short, they are considered to offer no contribution to reproduction. The death rate caused by competition is accounted for by μN for each group (S, H and I). Anderson *et al.* (1981) summarized the relevant values for these parameters, as shown in Table 10.1.

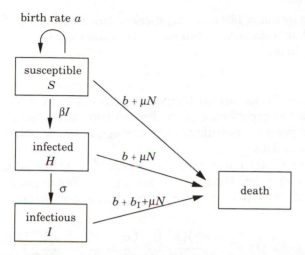

Fig. 10.2 Flow diagram of process of rabies virus infection. $N = S + H + I$; μN is the death rate due to the density effect. See Fig. 9.6 for other parameters.

We assume that the fox population before the introduction of rabies is maintained in a stable equilibrium, that is, carrying capacity K, determined by eqn (10.2). Thus, the density of the susceptible population at time $t = 0$ is set as

$$S(0) = K.$$

The criterion for determining whether rabies will spread or not when a few infective (i.e., latent and rabid) foxes enter this population can be obtained in a similar manner to that described at the end of the previous chapter (or by letting $S_0 \to K$, $b_H \to 0$, $b \to a(= b + \mu k)$, and $\gamma = 0$ in eqn (9.6)). Thus we have the condition for initial establishment of rabies in the following form:

$$K > K_t \equiv \frac{(a + \sigma)(a + b_I)}{\beta \sigma}. \tag{10.4}$$

Table 10.1 Values of demographic and epidemiological parameters of the red fox

Symbol	Meaning	Value
a	birth rate	1.0/year
b	death rate	0.5/year
$K = (a - b)/\mu$	fox carrying capacity	various
β	transmission coefficient	80 km^2/year
σ	$1/\sigma$ is the average latent period	13/year
b_I	death rate of rabid foxes	73/year

K_t, defined by the right-hand side, is called the 'threshold host density' for the spread of infection. Substituting the values of Table 10.1 into eqn (10.4), we obtain

$$K_t = 0.996 \text{ foxes per km}^2.$$

Indeed, when the fox density (carrying capacity K) was checked for those areas which had experienced the rabies epidemic, almost all showed values above 1 fox per km². Equation (10.4) thus agrees remarkably well with the observed field data.

When $K > K_t$, then, how will rabies spread through the population? It is easily shown that eqns (10.3) have a positive equilibrium point (S^*, H^*, I^*) when $K > K_t$ (whereas they have no positive equilibrium point when $K < K_t$):

$$I^* = \frac{(a-b)\{K\sigma\beta - (\sigma+a)(b_1+a)\}}{\beta\{K\sigma\beta - a(a-b)\}},$$

$$H^* = \frac{(b_1 + a - \beta I^*)I^*}{\sigma}, \tag{10.5}$$

$$S^* = K - H^* - I^* - \frac{\beta K I^*}{a-b}.$$

The total population density at this equilibrium is given by

$$N^* = S^* + H^* + I^* = \frac{K\{(\sigma+a)(b_1+a) - a(a-b)\}}{\sigma\beta K - a(a-b)}. \tag{10.6}$$

Moreover, when eqns (10.3) were numerically solved, we found that whenever a few infected foxes enter a fox population for which $K > K_t$, convergence to the equilibrium state of eqns (10.5) (or a limit cycle around the equilibrium) always occurs. Yet the manner in which S, H and I change over time varies greatly depending on the value of the carrying capacity K. Thus:

(1) if $K_t > K$, rabies does not become endemic;
(2) if $K_d > K > K_t$, S, H and I monotonically approach the equilibrium point (S^*, H^*, I^*);
(3) if $K_s > K > K_d$, they oscillate while approaching (S^*, H^*, I^*);
(4) if $K > K_s$, they approach a periodic solution (a limit cycle) that oscillates around the equilibrium point.

Here $K_t = 0.996$, $K_d = 1.006$ and $K_s = 8.57$. Figure 10.3 shows the temporal change in densities of susceptibles S and infectives $H + I$ for $K = 1, 2$ and 10, illustrating the above cases (2), (3) and (4), respectively. As shown here, the higher the carrying capacity K, the more violent is the oscillation

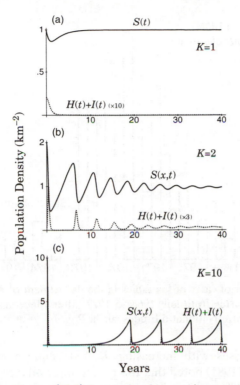

Fig. 10.3 Solutions of eqns (10.3) for varying fox carrying capacity K. S, H and I are fox densities of susceptible, infected and infectious individuals, respectively. (a) When $K = 1$, the population density of infectives, $H + I$, monotonically decreases to a very low level with time. (b) When $K = 2$, $H + I$ shows damping oscillation to approach an endemic level. (c) When $K = 10$, $H + I$ shows periodic oscillation with serrated profile. Parameter values are taken from Table 10.1.

of susceptible and infected fox populations, and the longer the oscillation lasts. For $1 < K < 4$, which is common in Europe, we see that the damped oscillation continues for several decades, with an oscillating cycle of 3–5 years. Indeed, as shown in Fig. 10.4, high incidences have been reported every 3–5 years in areas where rabies epidemics have occurred.

Figure 10.5(a) and (b) show, as functions of K, the density of susceptible foxes S^* (or average density in the case of limit cycles) and prevalence of rabies (i.e., the percentage of infected foxes in the total fox population, $y^* = (H^* + I^*)/N^*$) that are reached at equilibrium after a sufficient passage of time. As seen in Fig. 10.5(a), rabies will spread if the initial fox density exceeds K_t, but because the number of deaths increases with higher K, the level of survivors remains at about $S^* = 1$ fox per km^2 regardless of the K value. Meanwhile, the equilibrium prevalence of

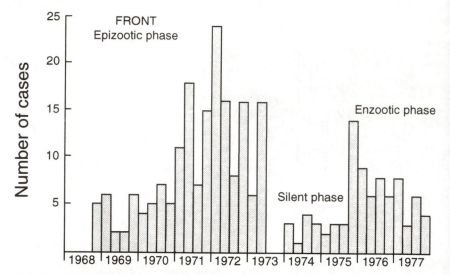

Fig. 10.4 Number of cases of fox rabies in the department of Ardennes in France recorded each quarter from late 1968 to 1977 (after Macdonald, 1980, based on data from the Centre National d'Études sur la Rage).

rabies y^* increases with increasing K, but remains within 4% or so. Anderson *et al.* (1981) noted that the reported prevalence of rabies is often in the range 3–7%, which is in close agreement with the theoretical value.

From Fig. 10.3, we see that when K is large, it takes a long time before equilibrium is reached, during which there are recurrent periods when the infected population density falls to extremely low levels. Thus, if the fox population lives within a closed, narrow range, the possibility exists that rabies will die out from stochastic effects (this was discussed in detail in the previous chapter with regard to measles). However, foxes are generally very active animals; young foxes usually establish their territories over 13 km away from their parents, while the adult fox normally covers a distance of 11–13 km each night. It is when such movements of foxes opens up new frontiers for rabies that it is able to spread its range. In the next section, we introduce a model which takes into account the spatial spread of rabies.

10.3 Spatial propagation

As we saw in Fig. 10.1, the range of the rabies epizootic in Europe is spreading yearly, and reports indicate that, as of 1990, it had crossed the River Loire in France (Murray and Seward, 1992). Figure 10.6(a) shows the reported incidence of rabies in the north-east region of France during

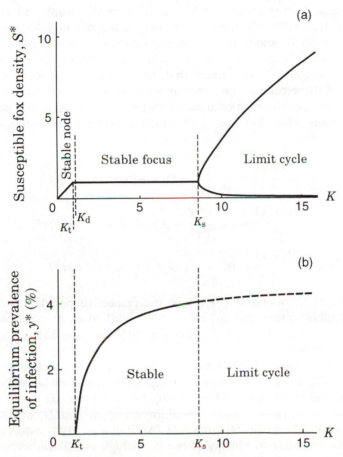

Fig. 10.5 (a) Susceptible fox density S^* at the equilibrium state as a function of fox carrying capacity K. Fox densities for the three stages, (S, H, I), approach stable node $(K, 0, 0)$ for $0 < K < K_t$, stable node (S^*, H^*, I^*) for $K_t < K < K_d$, stable focus (S^*, H^*, I^*) for $K_d < K < K_s$, and stable limit cycle around (S^*, H^*, I^*) for $K_s < K$, where $K_t = 0.996$, $K_d = 1.006$, $K_s = 8.57$. (S^*, H^*, I^*) is given by eqns (10.5) in the text. The two solid curves for $K_s > K$ indicate the upper and lower bounds of the limit cycle. (After Yachi *et al.*, 1989.) (b) Prevalence of infection at the equilibrium state, $y^* = (H^* + I^*)/(S^* + H^* + I^*)$, as a function of fox carrying capacity K. Dashed line in the limit cycle region denotes the average prevalence of rabies throughout one cycle. Prevalence rises, tending to about 4%, as K increases. (After Anderson *et al.*, 1981.)

1976. The range front is expanding in the southwesterly direction, with the highest reported frequencies at the advancing front. The number of dots falls drastically a small distance behind the front, then gradually begins rising again as one moves further away from the front and enters an epidemic area again. Macdonald (1980) expressed this in a diagram, shown in Fig. 10.6(b). It demonstrates that the rabies epidemic is most active at the advancing range front, behind which alternately appear, at cyclic intervals of 100–300 km, zones that have become quiescent after the passage of the epidemic and zones undergoing a recurrent epidemic.

Below we examine such patterns of spatial propagation using the following equations, which are eqns (10.3) plus added terms that account for the diffusion of foxes:

$$\frac{\partial S(x,t)}{\partial t} = D_S \frac{\partial^2 S}{\partial x^2} + (a-b)S - \mu SN - \beta SI,$$

$$\frac{\partial H(x,t)}{\partial t} = D_H \frac{\partial^2 H}{\partial x^2} + \beta SI - (b + \sigma + \mu N)H, \qquad (10.7)$$

$$\frac{\partial I(x,t)}{\partial t} = D_I \frac{\partial^2 I}{\partial x^2} + \sigma H - (b + b_I + \mu N)I.$$

Here $S(x,t)$, $H(x,t)$ and $I(x,t)$ are the respective population densities of susceptible, latent and infectious (or rabid) foxes at position x and time t, and D_S, D_H and D_I are the diffusion coefficients for these respective groups.

As defined by eqn (3.7) in Chapter 3, the diffusion coefficient is given by $\langle x^2 \rangle / 4t$, that is, the mean square of the straight-line distance that a fox travels in unit time, divided by 4. Using this formula, and based on field data obtained by radar tracking, the diffusion coefficient D_I of a rabid fox is estimated to be about 50 km²/year (Andral *et al.*, 1982; Murray, 1986) (although according to Murray, it can be as high as 330 km²/year during active periods). Meanwhile, the diffusion of healthy, susceptible foxes takes place primarily by the dispersal of young foxes leaving their parents' den to establish new territories. Since young foxes normally disperse to a location 13 km or more from their birthplace (Macdonald, 1980), $\langle x^2 \rangle =$

▷

Fig. 10.6 (a) Reported cases of rabies in wild animals in France from January to December 1976. Infested area is divided into three different phases: the epizootic phase, the silent phase and the endemic phase. At the front, reported cases (indicated by dots) are densely distributed, while the area behind the front is in a silent phase with cases sparsely confirmed. In the region behind the silent phase, a second epidemic (endemic phase) appears (Macdonald, 1980, based on data from the Centre National d'Études sur la Rage). (b) Fox population density as a function of the passage of the rabies epizootic (redrawn from Macdonald, 1980). See text for details.

13^2 km^2, and since this happens once a year, $t = 1$, we obtain $D_S = 13^2/4 = 42$ km^2/year. Because the diffusion coefficients of susceptible and rabid foxes are roughly of the same order, for simplicity we assume in our analysis that $D_S = D_H = D_I = D$. In reality, determining the diffusion

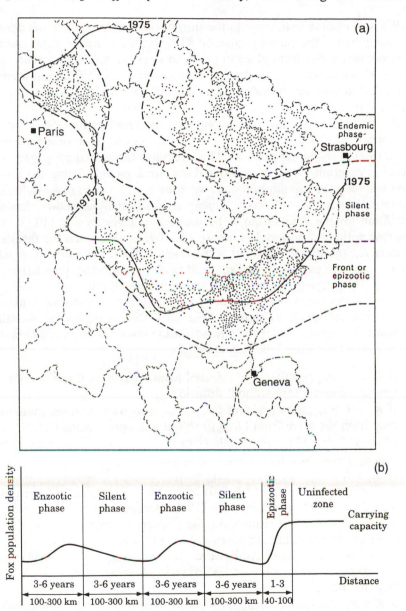

coefficient from field observation is a very difficult task, with estimates varying widely among workers; it is perhaps the least reliable of this model's parameters. As we shall see later, Murray *et al.* (1986) and Murray and Seward (1992) dealt with cases where D_I is much larger than D_S or D_H.

We now assume that, prior to the introduction of rabies, the fox density is maintained at the carrying capacity $K(x)$ specific to each locality. As the environment varies from place to place, so does the carrying capacity, and hence K is a function of position x. If we consider now a situation where a few infected foxes are introduced to a location ($x = 0$), in what manner will rabies spread?

We first introduce the results obtained by Yachi *et al.* (1989) for the case when $K(x) = K$ (constant), that is, a homogeneous environment with a constant carrying capacity, K. The condition for spatial propagation to occur is determined, as in previous discussions, by calculating the condition whereby the equilibrium state of eqns (10.7), $(S, H, I) = (K, 0, 0)$, is unstable against disturbances. Linear stability analysis shows that the condition for the equilibrium to be unstable is given by eqn (10.4), as in the case without diffusion. In other words, if K is lower than the threshold density K_t, the disease dies out without spreading. Conversely, if $K > K_t$, rabies will successfully establish itself and the infected population increases. By carrying out numerical computations for the latter case, we found that the range eventually develops into a propagating wave advancing at constant speed, as shown in Fig. 10.7. Depending on the K value, however, the spatial pattern of the advancing wave varies widely from the front to the tail in the following manner:

1. If $K_t > K$, the population of infected foxes dies out and rabies does not spread, as in the case without diffusion.
2. If $K_d > K > K_t$, the density of the propagating wave changes monotonically from the wave front $(K, 0, 0)$ to the tail, approaching (S^*, H^*, I^*), where (S^*, H^*, I^*) is the equilibrium state of eqns (10.5), for the case without diffusion (Fig. 10.7a).
3. If $K_s > K > K_d$, the propagating waves for S, H and I decay while oscillating from the front $(K, 0, 0)$ to the tail, converging to the equilibrium (S^*, H^*, I^*) given by eqns (10.5). Because the wave advances at constant speed while maintaining its pattern, to an observer at a fixed point successive epidemic waves appear to follow the original epidemic every few years. The peaks, however, become progressively smaller, and the disease eventually settles to an endemic state (S^*, H^*, I^*) (Fig. 10.7b).
4. If $K > K_s$, the spatial oscillations from front to tail make violent swings and span great distances; following the oscillations, there abruptly appear irregular fluctuations about the equilibrium point (Fig. 10.7c).

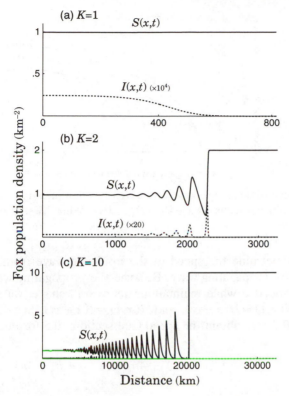

Fig. 10.7 Spatial patterns of propagating waves of rabies for various fox carrying capacities K. Rabid foxes are introduced locally in one-dimensional space, where susceptible foxes are distributed uniformly at a level equivalent to its carrying capacity K. The rabid foxes will establish a travelling wave with a monotonically increasing tail for $K_t < K < K_d$, oscillatory decaying tail for $K_d < K < K_s$ and irregularly oscillating tail for $K_s < K$. Typical profiles of travelling waves that move to the right are shown in (a) for $K = 1.0$, where $(S^*, H^*, I^*) = (0.996, 0.137 \times 10^{-3}, 0.24 \times 10^{-4})$ and $c = 3$ km/year, in (b) for $K = 2.0$, where $(S^*, H^*, I^*) = (0.975, 0.178 \times 10^{-1}, 0.314 \times 10^{-2})$ and $c = 46$ km/year, and in (c) for $K = 10$, where $(S^*, H^*, I^*) = (0.958, 0.319 \times 10^{-1}, 0.563 \times 10^{-2})$ and $c = 112$ km/year. $D_S = D_H = D_I = D$ is taken as 50 km²/year throughout, and other parameters are the same as in Table 10.1. (Modified from Yachi *et al.*, 1989.)

Here, $K_t = 0.996$, $K_d = 1.006$ and $K_s = 8.57$, the same values as those appearing on page 170 describing the dynamics of the rabies epidemic without the diffusion terms.

As seen here, the tail pattern varies widely depending on the K value. The range $1 < K < 4$, commonly seen in Europe, falls under category 3 above. Indeed, the patterns of Fig. 10.6(a) and (b) for France are of this type observed at a fixed point in time.

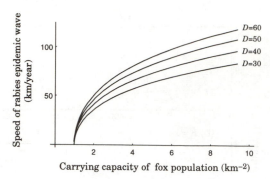

Fig. 10.8 Speed of rabies epizootic wave c as a function of carrying capacity K for various diffusion constants D ($= D_S = D_H = D_I$). When $K < K_t = 0.996$, rabies cannot expand.

We now determine the speed of the epidemic wave front, that is, the speed c of the propagating wave. Because the propagating wave advances at constant speed c while maintaining its overall shape, we set $S(x, t) = S(x - ct)$, $H(x, t) = H(x - ct)$, and $I(x, t) = I(x - ct)$. Substituting these into eqns (10.7), we obtain the speed (see section 10.5 for derivation):

$$c = \sqrt{2D\left\{\sqrt{(\sigma - b_1)^2 + 4\sigma\beta K} - (\sigma + b_1 + 2a)\right\}} . \qquad (10.8)$$

This shows that c is proportional to the square root of the diffusion coefficient D. Substituting the values of Table 10.1 into the above equation, we obtain the relation between c and K as shown in Fig. 10.8. When K is smaller than $K_t = 0.996$, rabies cannot take hold and so $c = 0$. For K larger than K_t, c increases monotonically with K, describing a curve that is convex upward. As mentioned before, the diffusion coefficient of red fox is thought to be around $D = 40$–50 km^2/year, so for $K = 1.5$ foxes per km^2, we have $c = 30$–33 km/year, and for $K = 4$ foxes per km^2, we have $c = 66$–73 km/year. Since $1 < K < 4$ in Europe, we can predict that the speed falls within this range, and indeed the observed average propagation speed of 30–60 km/year is in good agreement.

So far we have assumed a uniform fox carrying capacity, that is, a homogeneous environment infinitely expanding in one-dimensional space. In reality, however, environmental factors vary from place to place, and the number of fox territories will vary accordingly. Figure 10.9 shows a contour plot of fox carrying capacities in southern England, estimated from actual fox counts (Murray *et al.*, 1986). It shows that $K(x)$ indeed varies widely depending on the locality. Based on the mapped carrying capacity $K(x)$ of Fig. 10.9, Murray *et al.* (1986) carried out a large-scale

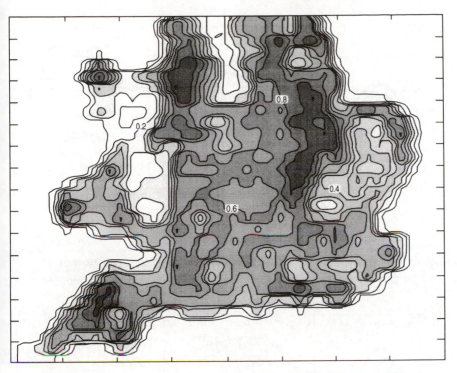

Fig. 10.9 Contour plot of fox densities in southern half of England. Values are scaled to lie between 0 and 1, with 1 corresponding to 2.4 adult foxes per km² in the springtime, or to an average of 4.6 foxes per km² throughout the year. (After Murray *et al.*, 1986, based on Macdonald's (1980) estimates.)

computer simulation in which eqns (10.7) were solved in a two-dimensional space that represented England's topography in order to predict the propagation route of rabies. Here the parameter values given in Table 10.1 and the diffusion coefficients $D_S = D_H = 0$ and $D_I = 200$ km²/year were used. Figure 10.10 shows the results. Assuming that rabies enters at Southampton in the south of England, contour plots of the density of rabid foxes are shown in one-year intervals. The rabies epidemic is seen to spread northward, advancing quickly where $K(x)$ is high and being slowed down where $K(x)$ is low, eventually reaching Manchester in about 3.5 years. We see how areas through which the epidemic has passed experience a quiescent period, but that seven years later a second epidemic of lower intensity breaks out at the original entry point and moves northward, taking a similar course.

There are several other approaches to derive the threshold density and speed of the rabies epidemic. In particular, the work of Mollison and

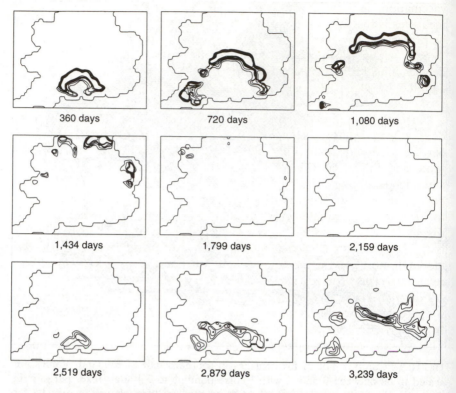

Fig. 10.10 Predicted epidemic front that starts from Southampton and moves through the southern part of England. Contour plots of rabid fox densities are obtained by solving eqns (10.7) with the local fox densities shown in Fig. 10.9. About 7 years after initial outbreak, another epidemic wave reappears at Southampton. Diffusion coefficients are $D_S = D_H = 0$, $D_I = 200$ km^2/year, and other parameter values are taken from Table 10.1. (After Murray *et al.*, 1986.)

Kuulasmaa (1985), incorporating stochastic dispersal, and that of van den Bosch *et al.* (1990), based on a mechanistic model as introduced in section 5.4, are most important.

10.4 Control of expansion

A major conclusion above was that spatial spread of rabies does not occur in areas where the host density K is lower than the threshold density K_t. Thus, to prevent the spread of rabies, one needs only to reduce the fox population in susceptible areas to below the threshold density. Although reducing the susceptible fox population can be done by culling or by vaccination, in view of the cost and labour involved, neither of these measures is realistic if they are to be undertaken over the entire region to be protected. An alternative, therefore, is to create a protective barrier, or 'break', within which the fox population is reduced below the threshold density, ahead of the advancing rabies front. This method has in fact been adopted by Denmark and some other countries in Europe (Wandeler *et al.*, 1988).

Currently the most economical way of reducing the susceptible fox density is to drop food containing vaccine and then wait for the foxes to eat the food and develop immunity. Not all foxes will take the prepared food, and the proportion of foxes that develop immunity, even if vast amounts of vaccine are dropped, cannot be expected to exceed 70% of the population at best. In areas with high fox density, therefore, it is difficult to lower the susceptible fox population to a level sufficiently below the threshold density, in which case a wider protective break is needed. What is the break width needed to prevent the invasion of rabies? Using the model discussed in the previous section, Murray *et al.* (1986) examined the relationship between fox density and break width.

Figure 10.11 depicts the behaviour of the propagating wave of rabies as it approaches and comes into contact with a break, computed numerically from eqns (10.7) for the case that $D_I = 200$ km^2/year and $D_S = D_H = 0$. The carrying capacity outside the break and the density of susceptible foxes within the break are set to be $K_{out} = 2$ foxes per km^2 and $K_{in} = 0.4$ fox per km^2, respectively. It can be seen that the peak of the propagating wave of infected foxes falls as it approaches the break. The density of infected foxes is extremely low inside the break, although the range front is seen to slowly advance forward. We denote as t_c the time at which infected foxes in the area falls below 0.5 fox per km^2. If the break width denoted by x_c is sufficiently large, the probability that an infected fox will reach the other side of the break can be considered to be negligibly small. (Mathematically speaking, the probability of an infected fox crossing the break will never fall to zero so long as the diffusion equations of (10.7) are used; if it is sufficiently small, however, we can consider that the infected fox will die before getting across, due to stochastic effects.) Assume now that, at $t = t_c$, the densities of infected foxes on both sides of the break have a ratio of δ, or

$$\delta\{H(0, t_c) + I(0, t_c)\} = H(x_c, t_c) + I(x_c, t_c) \qquad (10.9)$$

The smaller δ is chosen, the lower is the chance that the disease will succeed in crossing the break. Thus Murray *et al.* (1986) set $\delta = 10^{-4}$ and calculated the break width x_c that satisfies eqn (10.9) for various values of the carrying capacity K_{in} in the break. The calculated break widths x_c are

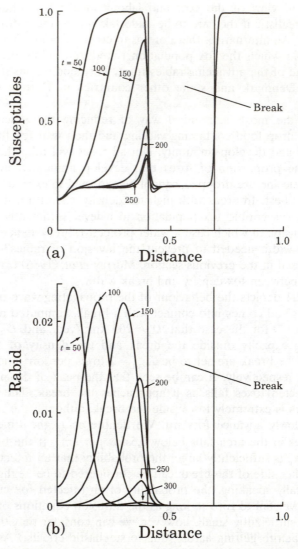

Fig. 10.11 Behaviour of epizootic front when it encounters a break. Equations (10.7) are solved for a carrying capacity outside the break of $K_{out} = 2$ fox per km^2 and a carrying capacity in the break of $K_{in} = 0.4$ fox per km^2. Other parameter values are the same as in Fig. 10.10. (a) Susceptible population density for a sequence of times. The density just outside the break remains slightly higher than elsewhere since few rabid foxes wander into this region from the right. (b) Corresponding rabid fox population densities. The wave approaches the break region, stops and dissipates. Times and distances are normalized values. (After Murray *et al.*, 1986.)

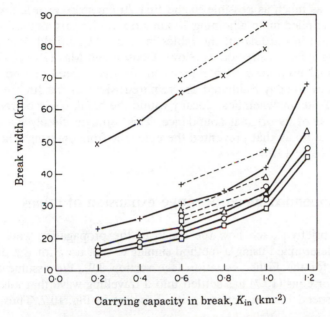

Fig. 10.12 Break width as a function of the carrying capacity in the break, K_{in}, for various values of diffusion coefficient for susceptible and infected foxes. Solid curves show cases for which the carrying capacity outside the break is fixed as $K_{out} = 2$ fox per km², while dashed curves correspond to $K_{out} = 4.6$ fox per km². $D_S = D_H$ (km²/year): □, 0; ○, 5; △, 20; +, 50; ×, 200. The diffusion coefficient for infectious foxes is fixed as $D_I = 200$ km²/year in all cases. Other parameters are taken from Table 10.1. See text for details.

plotted in Fig. 10.12 (squares). From this, we see that if $K_{out} = 2$ and $K_{in} = 0.6$ (70% reduction), when the vaccine is working effectively, a break width of 20 km is needed, and if $K_{in} = 1.2$ (40% reduction), when the vaccine has not spread widely, a break width of 50 km is needed.

Murray and Seward (1992) further extended the analysis to cases in which susceptible and latent foxes also undergo diffusion. The computation was performed for various values of diffusion coefficients, $D_S = D_H = 5$, 20, 50 or 200 km²/year, and the carrying capacity was set as $K_{out} = 2$ and 4.6 foxes per km², which correspond to the approximate carrying capacities for foxes in continental Europe and England, respectively. The results are plotted in Fig. 10.12 together with the previous result. It is seen that break widths are similar in all cases except when the diffusion coefficient is large. Moreover, the break widths for $K_{out} = 4.6$ are only about 5 km wider than those for $K_{out} = 2$. Denmark has adopted a two-layer break, where hunting and gassing fox dens are carried out to lower K_{in} as much as possible in the first 20 km strip, with less intensive measures applied in an adjoining 20 km strip. So far this method seems to be successful in warding off the rabies invasion (Macdonald, 1980).

By using a similar method as above, Dobson and May (1986a) discussed the potential usefulness of firebreaks to halt the spread of rinderpest in Africa. They roughly estimated the required width of the firebreak to be around 50–70 km, when host density within the break could be reduced to 20% or less of its original abundance. This estimate closely matches the width of the break that prevented the epidemic from entering Rhodesia in 1937–41 (Scott, 1970).

10.5 Appendix: speed of range expansion of rabies

Here we briefly sketch how the speed of the propagating wave of eqns (10.7) is determined using a method similar to that used for the propagating wave for the Fisher equation (see section 3.9). We assume that the solution for eqns (10.7) has settled into a travelling wave that advances at constant speed c while retaining its shape, as in Fig. 10.7. Thus, we can write

$$S(x,t) = S(x - ct) = S(\xi),$$
$$H(x,t) = H(x - ct) = H(\xi),$$
$$I(x,t) = I(x - ct) = I(\xi),$$

where $\xi = x - ct$, which, substituted into eqns (10.7), give

$$-cS' = DS'' + (a - b)S - \mu SN - \beta SI,$$
$$-cH' = DH'' + \beta SI - (b + \sigma + \mu N)H, \qquad (10.10)$$
$$-cI' = DI'' + \sigma H - (b + b_I + \mu N)I,$$

where a single prime (') denotes $d/d\xi$ and a double prime (") $d^2/d\xi^2$. Denoting $U = dS/d\xi$, $V = dH/d\xi$ and $W = dI/d\xi$, eqn (10.10) becomes the following set of differential equations:

$$S' = U,$$
$$H' = V,$$
$$I' = W,$$
$$U' = -\frac{c}{D}U - \frac{1}{D}((a - b)S - \mu SN - \beta SI),$$
$$V' = -\frac{c}{D}V - \frac{1}{D}(\beta SI - (b + \sigma + \mu N)H), \qquad (10.11)$$
$$W' = -\frac{c}{D}W - \frac{1}{D}(\sigma H - (b + b_I + \mu N)I).$$

As seen in Fig. 10.7, far ahead of the travelling wave front to the right ($\xi \to \infty$), we have

$$(S(\infty), H(\infty), I(\infty), U(\infty), V(\infty), W(\infty)) = (K, 0, 0, 0, 0, 0). \quad (10.12)$$

Meanwhile, far behind the wave front to the left ($\xi \to -\infty$), the following must hold:

$$(S(-\infty), H(-\infty), I(-\infty), U(-\infty), V(-\infty), W(-\infty))$$
$$= (S^*, H^*, I^*, 0, 0, 0), \quad (10.13)$$

where (S^*, H^*, I^*) is the equilibrium point given by eqns (10.5). Equations (10.12) and (10.13) are both equilibrium points of eqns (10.11). So the orbit of the travelling wave solution of eqns (10.11) must connect these two equilibrium points while maintaining positive values of S, H and I. We thus determine the condition that gives a non-negative travelling wave solution in the vicinity of $(K, 0, 0, 0, 0, 0)$. To this end, we take the linear approximation of eqns (10.11) in the vicinity of equilibrium point $(K, 0, 0, 0, 0, 0)$ and construct its Jacobian matrix. The eigenvalues of the Jacobian matrix are determined by the following equation:

$$(D\lambda^2 + c\lambda + a_1)(D\lambda^2 + c\lambda + a_2)(D\lambda^2 + c\lambda + a_3) = 0, \quad (10.14)$$

where

$$a_1 = -(a - b), \qquad a_2 = \frac{-p - \sqrt{q}}{2}, \qquad a_3 = \frac{-p + \sqrt{q}}{2},$$

$$p = \sigma + b_I + 2a, \qquad q = (\sigma - b_I)^2 + 4\sigma\beta K.$$

As the orbits of the travelling wave solutions approach eqn (10.12) as $\xi \to \infty$, they should be confined in the subspace spanned by eigenvectors that correspond to the eigenvalues with negative real parts. From the requirement that the H and I components of those eigenvectors are positive, we find that eigenvalues should be the solutions of

$$D\lambda^2 + c\lambda + a_3 = 0, \quad (10.15)$$

and furthermore consist only of real parts. Hence the following inequality must hold:

$$c^2 \geq 4Da_3.$$

While this means that the speed of the travelling wave must be at least $2\sqrt{Da_3}$, it has been confirmed from numerical computation that the speed of a stable travelling wave is in fact equal to this minimum value, or

$$c = 2\sqrt{Da_3}. \quad (10.16)$$

Substituting the expression for a_3 of eqn (10.14) into eqn (10.16) leads to eqn (10.8). For a more detailed analysis, see Yachi *et al.* (1989).

References

Note: asterisks denote publications from the SCOPE (Scientific Committee on Problems of the Environment) programme on the ecology of biological invasions (di Castri *et al.*, 1990; Drake *et al.*, 1989; Groves and Burdon, 1986; Groves and di Castri, 1991; Kornberg and Williamson, 1986; Mooney and Drake, 1986; Ramakrishnan, 1991).

Ali, S. W. and Cosner, C. (1995). Models for the effects of individual size and spatial scale on competition between species in heterogeneous environments. *Mathematical Biosciences* **127**, 45–76

Allee, W. C. (1938). *The Social Life of Animals*. W. W. Norton and Co., New York.

Ammerman, A. J. and Cavalli-Sforza, L. L. (1984). *The Neolithic Transition and the Genetics of Populations in Europe*. Princeton University Press, Princeton.

Anderson, R. M. and May, R. M. (1979). Population biology of infectious diseases: Part 1. *Nature* **280**, 361–367.

Anderson, R. M. and May, R. M. (1982). Directly transmitted infectious diseases: Control by vaccination. *Science* **215**, 1053–1060.

Anderson, R. M. and May, R. M. (1986). The invasion, persistence and spread of infectious diseases within animal and plant communities. *Philosophical Transactions of the Royal Society of London, Series B* **314**, 533–570.

Anderson, R. M. and May, R. M. (1991). *Infectious Diseases of Humans: Dynamics and Control*. Oxford Science Publications.

Anderson, R. M., Jackson, H. C., May, R. M. and Smith, A. M. (1981). Population dynamics of fox rabies in Europe. *Nature* **289**, 765–771.

Andow, D., Kareiva, P., Levin, S. and Okubo, A. (1990). Spread of invading organisms. *Landscape Ecology* **4**, 177–188.

Andow, D., Kareiva, P., Levin, S. and Okubo, A. (1993). Spread of invading organisms: Patterns of spread. In *Evolution of insect pests: The pattern of variations* (ed. K. C. Kim), pp. 219–242. Wiley, New York.

Andral, L., Artois, M., Aubert, M. F. A. and Blancou, J. (1982). Radio-pistage de renards enragés. *Comparative Immunology, Microbiology and Infectious Diseases* **5**, 284–291.

Aoki, K., Shida, M. and Shigesada, N. (1996). Travelling wave solutions for the spread of farmers into a region occupied by hunter-gatherers. *Theoretical Population Biology* **50**, 1–17.

Armstrong, R. A. (1976). Fugitive species: experiments with fungi and some theoretical considerations. *Ecology* **57**, 953–963.

Aronson, D. G. (1980). Density-dependent interaction-diffusion systems. In *Dynamics and Modelling of Reactive Systems* (ed. W. E. Stewart, W. H. Ray, and C. C. Conley), pp. 161–176. Academic Press, New York.

Aronson, D. G. and Weinberger, H. F. (1975). Nonlinear diffuse in population genetics, combustion, and nerve pulse propagation. In *Partial differential equations and related topics* (ed. E. A. Goldstein), Lecture Notes in Mathematics, vol. 446, pp. 5–49. Springer-Verlag, Berlin.

Aronson, D. G. and Weinberger, H. F. (1978). Multi-dimensional nonlinear diffusion arising in population genetics. *Advances in Mathematics* **30**, 33–76.

Bacon, P. J. (ed.) (1985). *Population Dynamics of Rabies in Wildlife*. Academic Press, London.

Bailey, N. T. J. (1975). *The Mathematical Theory of Infectious Diseases and its Appellations*. Charles Griffin, London.

Baker, H. G. (1962). Weeds—native and introduced. *Journal of the California Horticultural Society* **23**, 97–104.

Baker, H. G. (1974). The evolution of weeds. *Annual Review of Ecology and Systematics* **5**, 1–24.

Baker, H. G. (1986). Patterns of plant invasion in North America. In *Ecology of Biological Invasions of North America and Hawaii* (ed. H. A. Mooney and J. A. Drake), pp. 44–57. Springer-Verlag, Berlin.

Banks, R. B. (1994). *Growth and Diffusion Phenomena*. Springer-Verlag, Berlin.

Barclay, H. and Mackauer, M. (1980). The sterile insect release method for pest control: a density-dependent model. *Environmental Entomology* **9**, 810–817.

Bartlett, M. S. (1957). Measles periodicity and community size. *Journal of the Royal Statistical Society* **A120**, 48–70.

Baumhover, A. H., Graham, A. J., Bitter, B. A., Hopkins, D. E., New, W. D., Dudley, F. H. and Bushland, R. C. (1955). Screwworm control through release of sterilized flies. *Journal of Economic Entomology* **48**, 462–466.

Bazzaz, F. A. (1986). Life history of colonizing plants: Some demographic, genetic, and physiological features. In *Ecology of Biological Invasions of North America and Hawaii* (ed. H. A. Mooney and J. A. Drake), pp. 96–110. Springer-Verlag, Berlin.

Beddington, J. R., Free, C. A. and Lawton, J. H. (1975). Dynamic complexity in predator–prey models framed in difference equations. *Nature* **255**, 58–60.

Bennett, K. D. (1983). Postglacial population expansion of forest trees in Norfolk, U.K. *Nature* **303**, 164–167.

Bennett, K. D. (1986). The rate of spread and population increase of forest trees during the postglacial. *Philosophical Transactions of the Royal Society of London, Series B* **314**. 523–531.

Berg, H. C. (1983). *Random Walks in Biology*. Princeton University Press.

Bramson, M. (1973). *Convergence of Solutions of the Kolmogorov Equation to Travelling Waves*. AMS Memoirs, No. 285, vol. 44, American Mathematical Society, Providence, RI.

Britton, N. F. (1986). *Reaction-Diffusion Equations and Their Applications to Biology*. Academic Press, London.

Brown, D. and Rothery, P. (1993). *Models in Biology: Mathematics, Statistics and Computing*. Wiley, New York.

Brown, J. H. (1989). Patterns, modes and extents of invasions by vertebrates. In *Biological Invasions: A Global Perspective* (ed. J. A. Drake, H. A Mooney, R. di Castri, R. H. Groves, F. J. Kruger, M. Rejmánek and M. Williamson). SCOPE 37, pp. 85–109. Wiley, New York.

Burgman, M. A., Ferson, S. and Akcakaya, H. R. (1993). *Risk Assessment in Conservation Biology*. Population and Community Biology Series. Chapman & Hall, London.

Cantrell, R. S. and Cosner, C. (1991). The effects of spatial heterogeneity in population dynamics. *Journal of Mathematical Biology* **29**, 315–338.

Capasso, V. (1993). *Mathematical Structures of Epidemic Systems*. Lecture Notes in Biomathematics 97. Springer-Verlag, Berlin.

Carslaw, H. S. and Jaeger, J. C. (1959). *Conduction of Heat in Solids*, 2nd edn. Clarendon Press, Oxford.

Case, T. J. (1991). Invasion resistance, species build-up and community collapse in metapopulation models with interspecies competition. *Biological Journal of the Linnean Society* **42**, 239–266.

Castillo-Chavez, C. (ed.) (1989). *Mathematical and Statistical Approaches to AIDS Epidemiology*. Lecture Notes in Biomathematics, vol. 83. Springer-Verlag, Berlin.

Caswell, H. (1978). Predator mediated coexistence: A nonequilibrium model. *American Naturalist* **112**, 127–154.

Caswell, H. (1989). *Matrix Population Models*. Sinauer Associations, Inc. Publishers, Sunderland, MA.

Caswell, H. and Cohen, J. E. (1991). Disturbance, interspecific interaction and diversity in metapopulations. *Biological Journal of the Linnean Society* **42**, 193–218.

Caswell, H. and Etter, R. J. (1992). Ecological interactions in patchy environments: from patch-occupancy models to cellular automata. In *Patch Dynamics* (ed. S. A. Levin, T. M. Powell and J. H. Steele), pp. 93–109. Springer, New York.

Caughley, G. (1970). Liberation, dispersal and distribution of Himalayan thar (*Hemitragus jemlahicus*) in New Zealand. *New Zealand Journal of Science* **13**, 200–239.

Caughley, G., Grice, D., Barker, R. and Brown, B. (1988). The edge of range. *Journal of Animal Ecology* **57**, 771–785.

Chandrasekhar, S. (1943). Stochastic problems in physics and astronomy. *Reviews of Modern Physics* **15**, 1–89.

Chesson, P. L. and Warner, R. R. (1981). Environmental variability promotes coexistence on lottery competitive systems. *American Naturalist* **117**, 923–943.

Clarke, C. M. H. (1971). Liberations and dispersal of red deer in northern South Island districts. *New Zealand Journal of Science* **1**, 194–207.

Cliff, A. P., Haggett, P., Ord, J. K. and Versey, G. R. (1981). *Spatial Diffusion: An Historical Geography of Epidemics in an Island Community*. Cambridge University Press.

Coddington, E. and Levinson, N. (1972). *Theory of Ordinary Differential Equations*. McGraw-Hill, New York.

Colasanti, R. L. and Grime, J. P. (1993). Resource dynamics and vegetation processes: a deterministic model using two-dimensional cellular automata. *Functional Ecology* **7**, 169–176.

Comins, H. N., Hassell, M. P., and May, R. M. (1992). The spatial dynamics of host–parasitoid systems. *Journal of Animal Ecology* **61**, 735–748.

Connell, J. H. (1978). Diversity in tropical rainforests and coral reefs. *Science* **199**, 1302–1310.

Cox, C. B. and Moore, P. D. (1993). *Biogeography. An Ecological and Evolutionary Approach*, 5th edn. Blackwell, Oxford.

Crawley, M. J. (1986). The population biology of invaders. *Philosophical Transactions of the Royal Society of London, Series B* **314**, 711–731.

Crawley, M. J. (1987). What makes a community invasible? In *Colonization, Succession and Stability* (ed. A. J. Gray, M. J. Crawley and P. J. Edwards), pp. 429–453. Blackwell, Oxford.

Crosby, A. W. (1986). *Ecological Imperialism. The Biological Expansion of Europe*, 900–1900. Cambridge University Press.

Daniels, H. S. (1975). The deterministic spread of a simple epidemic. In *Perspective in Probability and Statistics*: Papers in honour of M. S. Bartlett on the occasion of his sixty-fifth birthday (ed. J. Gani), pp. 373–386. Academic Press, London.

Davis, M. B. (1976). Pleistocene biogeography of temperate deciduous forest. *Geoscience and Man* **13**, 13–26.

Davis, M. B. (1981). Quaternary history and the stability of forest communities. In *Forest Succession: Concepts and Applications* (ed. D. C. West, H. H. Shugart and D. B. Botkin), pp. 132–153. Springer-Verlag, New York.

DeAngelis, D. L., Waterhouse, J. C., Post, W. M. and O'Neill, R. V. (1985). Ecological modelling and disturbance evaluation. *Ecological Modelling* **29**, 399–419.

Denslow, J. S. (1987). Tropical rainforest gaps and tree species diversity. *Annual Review of Ecology and Systematics* **18**, 431–451.

Dexter, F., Banks, H. T. and Webb, T., III (1987). Modeling Holocene changes in the location and abundance of beech populations in eastern North America. *Review of Palaeobotany and Palynology* **50**, 273–292.

Di Castri, F. (1989). History of biological invasions with special emphasis on the old world. In *Biological Invasions: A Global Perspective* (ed. J. A. Drake, H. A. Mooney, F. di Castri, R. H. Groves, F. J. Kruger, M. Rejmánek and M. Williamson), SCOPE 37, pp. 31–55. Wiley, New York.

* Di Castri, F., Hansen, A. J. and Debussche, M. (eds). (1990). *Biological Invasions in Europe and the Mediterranean Basin*. Kluwer, Dordrecht.

Diamond, J. and Case, T. J. (1986). Overview: introductions, extinctions, exterminations, and invasions. In *Community Ecology* (ed. J. Diamond and T. J. Case), pp. 65–79. Harper and Row, New York.

Diamond, J. M. and May, R. M. (1981). Island biogeography and the design of nature reserve. In *Theoretical Ecology: Principles and Applications* (ed. R. M. May), pp. 228–252. Blackwell, Oxford.

Doak, D. F., Marino, P. L. and Kareiva, P. M. (1992). Spatial scale mediates the influence of habitat fragmentation on dispersal success: implications for conservation. *Theoretical Population Biology* **41**, 315–336.

Dobson, A. P. and May, R. M. (1986a). Disease and conservation. In *Conservation Biology* (ed. M. E. Soule), pp. 345–365. Sinauer Associates, Inc., Sunderland, MA.

Dobson, A. P. and May, R. M. (1986b). Patterns of invasions by pathogens and parasites. In *Ecology of Biological Invasions of North America and Hawaii* (ed. H. A. Mooney and J. A. Drake), pp. 58–76. Springer-Verlag, Berlin.

* Drake, J. A., Mooney, H. A., di Castri, F., Groves, R. H., Kruger, F. J., Rejmánek, M. and Williamson, M. (eds) (1989). *Biological Invasions: A Global Perspective*, SCOPE 37. Wiley, New York.

Dunbar, S. R. (1983). Traveling wave solutions of diffusive Lotka–Volterra equations. *Journal of Mathematical Biology* **17**, 11–32.

Dunbar, S. R. (1984). Traveling wave solutions of diffusive Lotka–Volterra equations: a heteroclinic connection in R^4. *Transactions of the American Mathematical Society* **286**, 557–594.

Durrett, R. and Levin, S. (1994). Stochastic spatial models: a user's guide to ecological applications. *Philosophical Transactions of the Royal Society of London, Series B* **343**, 329–350.

Ehrlich, P. R. (1989). Attributes of invaders and the invading processes: vertebrates. In *Biological Invasions: A Global Perspective* (ed. J. A. Drake, H. A. Mooney, F. di Castri, R. H. Groves, F. J. Kruger, M. Rejmánek, and M. Williamson), SCOPE 37, pp. 315–328. Wiley, New York.

Einstein, A. (1905). Über die von der molekularkinetischen Theorie der Wärme geforderte Bewegung von in ruhenden Flüssigkeiten suspendierten Teilchen. *Annalen der Physik* **17**, 549–560.

Elkinton, J. S. and Liebhold, A. M. (1990). Population dynamics of gypsy moth in North America. *Annual Review of Entomology* **35**, 571–596.

Elton, C. S. (1958). *The Ecology of Invasion by Animals and Plants*. Methuen, London.

Etter, R. J. and Caswell, H. (1994). The advantages of dispersal in a patchy environment: effects of disturbance in a cellular automaton model. In *Reproduction, Larval Biology and Recruitment in the Deep-Sea Benthos* (ed. K. J. Eckelbarger and C. M. Young), pp. 285–305. Columbia University Press.

Fife, P. C. (1979). *Mathematical Aspects of Reacting and Diffusing Systems*. Lecture Notes in Biomathematics, vol. 28. Springer-Verlag, Berlin.

Fisher, R. A. (1937). The wave of advance of advantageous genes. *Annals of Eugenics* **7**, 255–369.

Fourier, J. (1822). *Analytical Theory of Heat*. Dover, New York.

Frank, V. F. and Härle, A. (1964). Die Entwicklung des Bisam-Befalls (*Ondatra zibethica*) in der Bundesrepublik Deutschland von 1957 bis 1963. *Nachrichtenblatt des Deutschen Pflanzenschutzdienstes* **16**, 145–147.

Gilpin, M.. E. (1988). A comment on Quinn and Hastings: Extinction in subdivided habitat. *Conservation Biology* **2**, 290–293.

Gilpin, M. E. and Diamond, J. M. (1980). Subdivision of nature reserves and the maintenance of species diversity. *Nature* **285**, 567–568.

Gilpin, M. and Hanski, I. (eds) (1991). *Metapopulation Dynamics: Empirical and Theoretical Investigations*. Academic Press, New York.

Goel, N. S. and Richter-Dyn, N. (1974). *Stochastic Models in Biology*, Academic Press, New York.

*Groves, R. H. and Burdon, J. J. (1986). *Ecology of Biological Invasion: An Australian Perspective*. Australian Academy of Science, Canberra.

*Groves, R. H. and di Castri, F. (eds) (1991). *Biogeography of Mediterranean Invasions*. Cambridge University Press.

Gyllenberg, M. and Hanski, I. (1992). Single-species metapopulation dynamics: a structured model. *Theoretical Population Biology* **42**, 35–61.

Hallam, T. G. and Levin, S. A. (eds) (1986). *Mathematical Ecology: An Introduction*. Springer-Verlag, New York.

Hamilton, W. D. (1980). Sex versus non-sex versus parasite. *Oikos* **35**, 282–290.

Hamilton, W. D. and Zuk, M. (1982). Heritable true fitness and bright birds: a role for parasites. *Science* **218**, 384–387.

Hamilton, W. D., Axelrod, R. and Tanese, R. (1990). Sexual reproduction as an adaptation to resist parasites (a review). *Proceedings of the National Academy of Sciences of the USA* **87**, 3566–3573.

Hanski, I. (1983). Coexistence of competitors in patch environment. *Ecology* **64**, 493–500.

Harada, Y. and Iwasa, Y. (1994). Lattice population dynamics for plants with dispersing seeds and vegetative propagation. *Researches on Population Ecology* **36**, 237–249.

Harada, Y., Ezoe, H., Iwasa, Y., Matsuda, H. and Sato, K. (1995). Population persistence and spatially limited social interaction. *Theoretical Population Biology* **48**, 65–91.

Harrison, S. (1994). Metapopulations and conservation. In *Large-Scale Ecology and Conservation Biology* (ed. P. J. Edwards, R. M. May and R. T. Webb), pp. 111–128. Blackwell, Oxford.

Hassell, M. P., Comins, H. N. and May, R. M. (1991). Spatial structure and chaos in insect population dynamics. *Nature* **353**, 255–258.

Hastings, A. (1980). Disturbance, coexistence, history, and competition for space. *Theoretical Population Biology* **18**, 3363–3373.

Hastings, A. (1991). Structured models of metapopulation dynamics. In *Metapopulation Dynamics: Empirical and Theoretical Investigations*. (ed. M. Gilpin and I. Hanski), pp. 57–71. Academic Press, New York.

Hastings, A. (1994). Conservation and spatial structure: theoretical approaches. In *Frontiers in Mathematical Biology* (ed. S. A. Levin), Lecture Notes in Biomathematics, vol. 100, pp. 494–503. Springer-Verlag, Berlin.

Hengeveld, R. (1989). *Dynamics of Biological Invasions*. Chapman and Hall, London.

Hengeveld, R. (1994). Small-step invasion research. *Trends in Ecology and Evolution* **9**, 339–342.

Hethcote, H. W. (1994). A thousand and one epidemic models. In *Frontiers in Mathematical Biology*, Lecture Notes in Biomathematics, vol. 100 (ed. S. A. Levin), pp. 504–515. Springer-Verlag, Berlin.

Hethcote, H. W. and Van Ark, J. W. (1987). Epidemiological models with heterogeneous populations: Proportionate mixing, parameter estimation and immunization programs. *Mathematical Biosciences* **84**, 85–118.

Hethcote, H. W. and Yorke, J. A. (1984). *Gonorrhea Transmission Dynamics and Control*, Lecture Notes in Biomathematics, vol. 56. Springer-Verlag, Berlin.

Heywood, V. H. (1989). Patterns, extents and modes of invasions by terrestrial plants. In *Biological Invasions: A Global Perspective* (ed. J. A. Drake, H. A. Mooney, F. di Castri, R. H. Groves, F. J. Kruger, M. Rejmánek and M. Williamson), SCOPE 37, pp. 31–55. Wiley, New York.

Higgs, A. J. and Usher, M. B. (1980). Should nature reserves be large or small? *Nature* **285**, 568–569.

Hoffman, M. (1958). Die Bisamratte. Akademische Verlagsgesellschaft, Leipzig.

Holmes, E. E. (1993). Are diffusion models too simple? A comparison with telegraph models of invasion. *American Naturalist* **142**, 779–795.

Holmes, E. E., Lewis, M. A., Banks, J. E. and Veit, R. R. (1994). Partial differential equations in ecology: Spatial interactions and population dynamics. *Ecology* **75**, 17–29.

Hoppensteadt, F. C. (1982). *Mathematical Methods of Population Biology*. Cambridge University Press.

Hosono, Y. (1989). Singular perturbation analysis of travelling waves for diffusive Lotka–Volterra competition models. In *Numerical and Applied Mathematics* (ed. C. Brezinski), pp. 689–692. J. C. Baltzer AG, Scientific Publishing Co., Basel, Switzerland.

Huston, M. (1979). A general hypothesis of species diversity. *American Naturalist* **113**, 81–101.

Isenmann, P. (1990). Some recent bird invasions in Europe and the Mediterranean Basin. In *Biological Invasions in Europe and the Mediterranean Basin* (ed. F. di Castri, A. H. Hansen and M. Debussche), pp. 245–262. Kluwer, Dordrecht.

Ito, Y. (1977). A model of sterile insect release for eradication of the melon fly, *Dacus cucurbitae* Coquillet. *Applied Entomology and Zoology* **12**, 303–312.

Iwahashi, O. (1977). Eradication of the melon fly, *Dacus cucurbitae*, from Kume Is., Okinawa with the sterile insect release method. *Researches on Population Ecology* **19**, 87–98.

Iwasa, Y. and Mochizuki, H. (1988). Probability of population extinction accompanying a temporary decrease of population size. *Researches on Population Ecology* **30**, 145–164.

Iwata, K., Kawasaki, K. and Shigesada, N. (1996). Dynamical model of the size distribution of multiple metastatic tumors (submitted).

Iwata, T. (1979). Invasion of the rice water weevil, *Lissorhoptrus oryzae* Kuschel, into Japan, spread of its distribution and abstract of the research experiments conducted in Japan. *Japan Pesticide Information* **367**, 14–21.

Jacobson, G. L., Webb, T. and Grimm, E. C. (1987). Patterns and rates of change during the deglaciation of eastern North America. In *The Geology of North America*, vol. K-3 (ed. W. F. Ruddiman and H. E. Wright), pp. 277–288. Geological Society of America, New York.

Jones, R., Gilbert, N., Guppy, M. and Nealis, V. (1980). Long distance movement of *Pieris rapae*. *Journal of Animal Ecology* **49**, 629–642.

Kakehashi, M. and Yoshinaga, F. (1992). Evolution of airborne infectious diseases according to changes in characteristics of the host population. *Ecological Research* **7**, 235–243.

Kallen, A., Arcuri, P. and Murray, J. D. (1985). A simple model for the spatial spread and control of rabies. *Journal of Theoretical Biology* **116**, 377–393.

Kametaka, Y. (1976). On the nonlinear diffusion equation of Kolmogorov–Petrovskii–Piscunov type. *Osaka Journal of Mathematics* **13**, 11–66.

Kan-on, Y. (1995). Parameter dependence of propagation speed of travelling waves for competition-diffusion equations. *SIAM Journal of Mathematical Analysis* **26**, 340–363.

Kareiva, P. M. (1982). Experimental and mathematical analyses of herbivore movement: Quantifying the influence of plant spacing and quality of foraging discrimination. *Ecological Monographs* **52**, 261–282.

Kareiva, P. M. (1983). Local movement in herbivorous insects: applying a passive diffusion model to mark–recapture field experiments. *Oecologia* **57**, 322–327.

Kareiva, P. M. (1987). Habitat fragmentation and the stability of predator–prey interactions. *Nature* **321**, 388–391.

Kareiva, P. M. and Shigesada, N. (1983). Analyzing insect movement as a correlated random walk. *Oecologia* **56**, 234–238.

Kareiva, P. and Wennergren, U. (1995). Connecting landscape patterns to ecosystem and population processes. *Nature* **373**, 299–302.

Kawasaki, K., Osawa, N., Takasu, F., Casewell, H. and Shigesada, N. (1996). Effect of stochastic dispersal on the rate of spatial propagation (submitted).

Kermack, W. O. and McKendrick, G. A. (1927). A contribution to the mathematical theory of epidemics. *Proceedings of the Royal Society of London, Series A* **115**, 700–721.

Kessel, B. (1953). Distribution and migration of the European starling in North America. *Condor* **55**, 49–67.

Kierstead, H. and Slobodkin, L. B. (1953). The size of water masses containing plankton bloom. *Journal of Marine Research* **12**, 141–147.

Kiritani, K. (1984). Colonizing insects (colonist as a member of native insect community). *The Insectarium* **9**, 248–262.

Kohyama, T. and Shigesada, N. (1996). A size-distribution-based model of forest dynamics over a thermal gradient along latitude. *Vegetatio* **124**, 117–126.

Kolmogorov, A., Petrovsky, N. and Picoounov, N. S. (1937). A study of the equation of diffusion with increase in the quantity of matter, and its application to a biological problem. *Moscow University Bulletin of Mathematics* **1**, 1–25.

*Kornberg, H. and Williamson, M. H. (eds) (1986). *Quantitative Aspects of the Ecology of Biological Invasions*. *Philosophical Transactions of the Royal Society of London, Series B*, vol. 314.

Kubo, T., Iwasa, Y. and Furumoto, N. (1996). Forest spatial dynamics with gap expansion: Total gap area and gap size distribution. *Journal of Theoretical Biology* **180**, 229–246.

Lande, R. (1987). Extinction thresholds in demographic models of territorial populations. *American Naturalist* **130**, 624–645.

Lande, R. (1988). Genetics and demography in biological conservation. *Science* **241**, 1455–1460.

Langer, W. L. (1964). The black death. *Scientific American* (February), 114–121.

Lawton, J. H. (1995). Population dynamic principles. In *Extinction Rates* (ed. J. H. Lawton and R. M. May), pp. 147–163. Oxford University Press.

Lawton, J. H. and Brown, K. C. (1986). The population and community ecology of invading insects. *Philosophical Transactions of the Royal Society of London, Series B* **314**, 607–617.

Lawton, J. H. and May, R. M. (eds) (1995). *Extinction Rates*. Oxford University Press.

Lawton, J. H., Nee, S., Letcher, A. J. and Harvey, P. H. (1994). Animal distributions: patterns and processes. In *Large-scale Ecology and Conservation Biology* (ed. P. J. Edwards, R. M. May and R. T. Webb), pp. 41–58. Blackwell, Oxford.

Leigh, E. G., Jr. (1981). The average lifetime of a population in a varying environment. *Journal of Theoretical Biology* **90**, 213–239.

Levin, S. A. (1989). Analysis of risk for invasions and control programs. In *Biological Invasions: A Global Perspective* (ed. J. A. Drake, H. A. Mooney, F. di Castri, R. H. Groves, F. J. Kruger, M. Rejmánek and M. Williamson), SCOPE 37, pp. 425–435. Wiley, New York.

Levin, S. A. (1992). The problem of pattern and scale in ecology. *Ecology* **73**, 1943–1967.

Levin, S. A. and Buttel, L. (1987). Measures of patchiness in ecological system. Report no. ERC-130. Ecosystems Research Center, Cornell University, Ithaca, NY.

Levin, S. A. and Paine, R. T. (1974). Disturbance, patch formation and community structure. *Proceedings of the National Academy of Sciences of the USA* **71**, 2744–2747.

Levin, S. and Pimentel, D. (1981). Selection of intermediate rates of increase in parasite–host systems. *American Naturalist* **117**, 308–315.

Levin, S. A., Moloney, K., Buttel, L. and Castillo-Chavez, C. (1989). Dynamical models of ecosystems and epidemics. *Future Generation Computer Systems* **5**, 265–274.

Lewis, M. A. and Kareiva, P. (1993). Allee dynamics and the spread of invading organisms. *Theoretical Population Biology* **43**, 141–158.

Lewis, M. A. and van den Driessche, P. (1993). Waves of extinction from sterile insect release. *Mathematical Biosciences* **116**, 221–247.

Liebhold, A. M. (1994). Gypsy moth in North America. USDA Forest Service Northeastern Forest Experimental Station. World Wide Web:http://gypsy.fsl.wvnet.edu/gmoth/isolated/isolated.html

Liebhold, A. M., Halverson, J. A. and Elmes, G. A. (1992). Gypsy moth invasion in North America: a quantitative analysis. *Journal of Biogeography* **19**, 513–520.

Lloyd, H. G. (1983). Past and present distribution of red and grey squirrels. *Mammal Review* **13**, 69–80.

Lodge, D. M. (1993). Biological invasions: lessons for ecology. *Trends in Ecology and Evolution* **8**, 133–137.

Lubina, J. A. and Levin, S. A. (1988). The spread of a reinvading species: range expansion in the California sea otter. *American Naturalist* **131**, 526–543.

MacArthur, R. H. and Wilson, E. O. (1967). *The Theory of Island Biogeography*. Princeton University Press.

Macdonald, D. W. (1980). *Rabies and Wildlife*. Oxford University Press.

Mack, R. (1970). The great African cattle plague epidemic of the 1980s. *Tropical Animal Health and Production* **2**, 210–219.

Mack, R. N. (1981). Invasion of *Bromus tectorum* L. into western north America: an ecological chronicle. *Agro-Ecosystems* **7**, 145–165.

Mack, R. N. (1985). Invading plants: their potential contribution to population biology. In *Studies on Plant Demography* (ed. J. White), pp. 127–142. Academic Press, London.

Mack, R. N. (1986). Alien plant invasion into the intermountain west: a case history. In *Ecology of Biological Invasions of North America and Hawaii* (ed. H. A. Mooney and J. A. Drake), Ecological Studies 58, pp. 191–213. Springer-Verlag, Berlin.

MacKinnon, K. (1978). Competition between red and grey squirrels. *Mammal Review* **8**, 185–190.

Magnus, W. and Winkler, S. (1966). Gene frequency clines in the presence of selection opposed by gene flow. *American Naturalist* **109**, 659–676.

Mason, C. J. and McManus, M. L. (1981). Larval dispersal of the gypsy moth. In *The Gyspy Moth: Research Toward Integrated Pest Management* (ed. C. C. Doane and M. L. McManus), Technical Bulletin 1584, pp. 161–202. USDA Forest Service.

May, R. M. (1978). The evolution of ecological systems. *Scientific American* **239**, 119–133.

May, R. M. (1985). Ecological aspects of disease and human populations. *American Zoologist* **25**, 441–450.

May, R. M. (1994a). The effects of spatial scale on ecological questions and answers, In *Large-scale Ecology and Conservation Biology* (ed. P. J. Edwards, R. M. May and R. T. Webb), pp. 1–17. Blackwell, Oxford.

May, R. M. (1994b). Spatial chaos and its role in ecology and evolution. In *Frontiers in Mathematical Biology*, Lecture Notes in Biomathematics, vol. 100, pp. 326–344. Springer-Verlag, Berlin.

May, R. M. and Anderson, R M (1983). Epidemiology and genetics in the coevolution of parasites and hosts. *Philosophical Transactions of the Royal Society of London, Series B* **219**, 281–313.

May, R. M. and Anderson, R. M. (1984). Spatial heterogeneity and the design of immunization programs. *Mathematical Biosciences* **72**, 83–111.

May, R. M. and Anderson, R. M. (1987). Transmission dynamics of HIV infection. *Nature* **326**, 137–142.

May, R. M. and Hassell, M. P. (1988). Population dynamics and biological control. *Philosophical Transactions of the Royal Society of London, Series B* **318**, 129–169.

McManus, M. L. and McIntyre, T. (1981). Introduction. In *The Gypsy Moth: Research Toward Integrated Pest Management* (ed. C. C. Doane and M. L. McManus), Technical Bulletin 1584, pp. 1–7. USDA Forest Service.

Metz, J. A. J. and Diekmann, O. (1986). *The Dynamics of Physiologically Structured Populations*. Springer-Verlag, New York.

Metz, J. A. J. and van den Bosch, F. (1995). Velocities of epidemic spread. In *Epidemic Models: Their Structure and Relation to Data* (ed. D. Mollison), pp. 150–186. Cambridge University Press.

Mollison, D. (1972). The rate of spatial propagation of simple epidemics. In *Proceedings of the Sixth Berkeley Symposium on Mathematical Statistics and Probability*, vol. 3, pp. 579–614.

Mollison, D. (1977). Spatial contact model for ecological and epidemic spread. *Journal of the Royal Statistical Society, Series B* **39**, 283–326.

Mollison, D. and Daniels, H. (1993). The 'deterministic simple epidemic' unmasked. *Mathematical Biosciences* **117**, 147–153.

Mollison, D. and Kuulasmaa, K. (1985). Spatial epidemic models: Theory and simulations. In *Population Dynamics of Rabies in Wildlife* (ed. P. J. Bacon), pp. 291–309. Academic Press, London.

Moloney, K. A., Levin, S. A. Chiariello, N. R. and Buttel, L. (1992). Pattern and scale in a serpentine grassland. *Theoretical Population Biology* **41**, 257–276.

* Mooney, H. A. and Drake, J. A. (eds) (1986). *Ecology of Biological Invasions of North America and Hawaii*. Ecological Studies, vol. 58. Springer-Verlag, New York.

Mooney, H. A. and Drake, J. A. (1989). Biological invasions: a SCOPE program, Overview. In *Biological Invasions: A Global Perspective* (ed. J. A. Drake, H. A. Mooney, F. di Castri, R. H. Groves, F. J. Kruger, M. Rejmánek, and M. Williamson), SCOPE 37, pp. 491–506. Wiley, New York.

Morisita, M. (1971). Measuring of habitat value by the 'environmental density' method. In *Statistical Ecology*, vol. 1 (ed. G. P. Patil, E. C. Pielou and W. E. Waters), pp. 379–401. Pennsylvania State University Press, University Park, PA.

Mundinger, P. C. and Hope, S. (1982). Expansion of the winter range of the House Finch: 1947–79. *American Birds* **36**, 347–353.

Murray, J. D. (1989). *Mathematical Biology*. Springer-Verlag, Berlin.

Murray, J. D. and Seward, W. L. (1992). On the spatial spread of rabies among foxes with immunity. *Journal of Theoretical Biology* **156**, 327–348.

Murray, J. D., Stanley, E. A. and Brown, D. L. (1986). On the spatial spread of rabies among foxes. *Proceedings of the Royal Society of London, Series B* **229**, 111–150.

Namba, T. and Mimura, M. (1980). Spatial distribution of competing populations. *Journal of Theoretical Biology* **87**, 795–814.

Nee, N. and May, R. M. (1992). Patch removal favours inferior competitors. *Journal of Animal Ecology* **61**, 37–40.

Newman, W. I. (1980). Some exact solutions to a nonlinear diffusion problem in population genetics and combustion. *Journal of Theoretical Biology* **85**, 325–334.

Nicholson, A. J. and Bailey, V. A. (1935). The balance of animal populations. Part 1. *Proceedings of the Zoological Society of London* **3**, 551–598.

Nobel, J. V. (1974). Geographic and temporal development of plagues. *Nature* **250**, 726–728.

Nowak, M. A., May, R. M. and Sigmund, K. (1995). The arithmetics of mutual help. *Scientific American* **272**(6), 50–55.

Okubo, A. (1980). *Diffusion and Ecological Problems: Mathematical Models*. Springer-Verlag, New York.

Okubo, A. (1988). Diffusion-type models for avian range expansion. In *Acta XIX Congress Internationalis Ornithologici 1* (ed. H. Quellet), National Museum of Natural Sciences, pp. 1038–1049. University of Ottawa Press.

Okubo, A. and Levin, S. A. (1989). A theoretical framework for data analysis of wind dispersal of seeds and pollen. *Ecology* **70**, 329–338.

Okubo, A., Maini, P. K., Williamson, M. H. and Murray, J. D. (1989). On the spatial spread of the grey squirrel in Britain. *Proceedings of the Royal Society of London, Series B* **238**, 113–125.

Pacala, S. W. and Silander, J. A., Jr (1985). Neighbourhood models of plant population dynamics. I. Single-species models of annuals. *American Naturalist* **125**, 385–411.

Pacala, S., Hassell, M. P. and May, R. M. (1990). Host–parasitoid associations in patchy environments. *Nature* **344**, 150–153.

Perry, C. C. (1955). *Gypsy moth appraisal program and proposed plan to prevent spread of the moths*, Technical Bulletin 1124. US Department of Agriculture.

Pickett, S. T. A. and White, P. S. (1985). The *Ecology of Natural Disturbance and Patch Dynamics*. Academic Press, New York.

Pimentel, D. (1986). Biological invasions of plants and animals in agriculture and forestry. In *Ecology of Biological Invasions of North America and Hawaii* (ed. H. A. Mooney and J. A. Drake), Ecological Studies 58, pp. 149–162. Springer-Verlag, New York.

Pimm, S. L. (1989). Theories of prediction success and impact of introduced species. In *Biological Invasions: A Global Perspective* (ed. J. A. Drake, H. A. Mooney, F. di Castri, R. H. Groves, F. J. Kruger, M. Rejmánek, and M. Williamson), SCOPE 37, pp. 351–367. Wiley, New York.

Piper, C. V. (1920). *Forage Plants and Their Culture*. MacMillan, New York.

Pyle, G. F. (1969). The diffusion of cholera in the United States in the nineteenth century. *Geographical Analysis* **1**, 59–75.

Pyle, G. F. (1982). Some observations on the geography of influenza diffusion. In *Influenza Models* (ed. P. Selby), pp. 213–223. MTP, Lancaster. UK.

Quinn, J. F. and Hastings, A. (1987). Extinction in subdivided habitats. *Conservation Biology* **1**, 198–208.

* Ramakrishnan, P. S. (ed.) (1991). *Ecology of Biological Invasion in the Tropics*. International Scientific Publications, New Delhi.

Rejmánek, M. (1984). Perturbation-dependent coexistence and species diversity in ecosystems. In *Stochastic Phenomena and Chaotic Behaviour in Complex Systems* (ed. P. Schuster), pp. 220–230. Springer-Verlag, Berlin.

Rejmánek, M. (1989). Invisibility of plant communities. In *Biological Invasions: A Global Perspective* (ed. J. A. Drake, H. A. Mooney, F. di Castri, R. H. Groves, F. J. Kruger, M. Rejmánek and M. Williamson), SCOPE 37, pp. 369–383. Wiley, New York.

Renshaw, E. (1991). *Modelling Biological Populations in Space and Time*. Cambridge University Press.

Reynolds, J. C. (1985). Details of the geographic replacement of the red squirrel (*Sciurus vulgaris*) by the grey squirrel (*Sciurus carolinensis*) in eastern England. *Journal of Animal Ecology* **54**, 149–162.

Rogers, D. J. and Randolph, S. E. (1986). Distribution and abundance of tsetse flies (*Glossina spp.*). *Journal of Animal Ecology* **55**, 1007–1025.

Roughgarden, J. (1979). *Theory of Population Genetics and Evolutionary Ecology: An Introduction*. Macmillan, New York.

Roughgarden, J. (1986). Predicting invastions and rates of spread. In *Ecology of Biological Invasions of North America and Hawaii* (ed. H. A. Mooney and J. A. Drake), pp. 179–188. Springer-Verlag, New York.

Sailer, R. J. (1983). History of insect introductions. In *Exotic plant pests and North American agriculture* (ed. C. L. Wilson and C. L. Graham), pp. 15–38. Academic Press, New York.

Sale, P. F. (1977). Maintenance of high diversity in coral reef fish communities. *American Naturalist* **111**, 337–359.

Sasaki, A. and Iwasa, Y. (1991). Optimal growth schedule of pathogens within a host: Switching between lytic and latent cycles. *Theoretical Population Biology* **39**, 201–239.

Sato, K., Matsuda, H. and Sasaki, A. (1994). Pathogen invasion and host extinction in lattice structured populations. *Journal of Mathematical Biology* **32**, 251–268.

Scott, G. R. (1970). Rinderpest. In *Infectious Diseases of Wild Mammals* (ed. J. W. Davis, L. H. Karsted and D. O. Trainer), pp. 20–35. Iowa State University Press.

Seno, H. (1989). The effect of a singular patch on population persistence in a multi-patch system. *Ecological Modelling* **43**, 271–286.

Seno, H. (1991). Predator's invasion into an isolated patch with spatially heterogeneous prey distribution. *Bulletin of Mathematical Biology* **53**, 557–577.

Shaffer, W. M. and Kot, M. (1985). Nearly one dimensional dynamics in an epidemic. *Journal of Theoretical Biology* **112**, 403–427.

Shigesada, N. (1992). *Mathematical Modeling for Biological Invasions*, UP Biology 92. University of Tokyo Press [in Japanese].

Shigesada, N. and Roughgarden, J. (1982). The role of rapid dispersal in the population dynamics of competition. *Theoretical Population Biology* **21**, 353–373.

Shigesada, N., Kawasaki, K. and Teramoto, E. (1979). Spatial segregation of interaction species. *Journal of Theoretical Biology* **79**, 83–99.

Shigesada, N., Kawasaki, K. and Teramoto, E. (1980). Spatial distribution of dispersing animals. *Journal of Mathematical Biology* **9**, 85–96.

Shigesada, N., Kawasaki, K. and Teramoto, E. (1986). Traveling periodic waves in heterogeneous environments. *Theoretical Population Biology* **30**, 143–160.

Shigesada, N., Kawasaki, K. and Teramoto, E. (1987). The speeds of traveling frontal waves in heterogeneous environments. In *Mathematical Topics in Population Biology, Morphogenesis and Neurosciences* (ed. E. Teramoto and M. Yamaguti), Lecture Notes in Biomathematics, vol. 71, pp. 88–97. Springer-Verlag, Berlin.

Shigesada, N., Kawasaki, K. and Takeda, Y. (1995). Modeling stratified diffusion in biological invasions. *American Naturalist* **146**, 229–251.

Simberloff, D. (1986). Introduced insects: A biogeographic and systematic perspective. In *Ecology of Biological Invasions of North America and Hawaii* (ed. H. A. Mooney and J. A. Drake), Ecological Studies, vol. 58, pp. 3–26. Springer-Verlag, New York.

Simberloff, D. (1989). Which insect introductions succeed and which fail? In *Biological Invasions: A Global Perspective* (ed. J. A. Drake, H. A. Mooney, F. di Castri, R. H. Groves, F. J. Kruger, M. Rejmánek and M. Williamson), SCOPE 37, pp. 61–75. Wiley, New York.

Simberloff, D. S. and Abele, L. G. (1976). Island biogeography theory and conservation practice. *Science* **191**, 285–286.

Simkin, T. and Fiske, R. S. (1983). *Krakatau 1883: The Volcanic Eruption and its Effects*. Smithsonian Institution Press, Washington DC.

Skellam, J. G. (1951). Random dispersal in theoretical populations. *Biometrika* **38**, 196–218.

Slatkin, M. (1974). Competition and regional existence. *Ecology* **55**, 128–134.

Smith, L. B. and Hadley, C. H. (1926). The Japanese beetle. Departmental Circular 363, pp. 1–66. US Department of Agriculture.

Soule, M. E. and Wilcox, B. A. (eds) (1980). *Conservation Biology: An Evolutionary Ecological Approach*. Sinauer, Sunderland, MA.

Sousa, W. P. (1979). Disturbance in marine intertidal boulder fields: the nonequilibrium maintenance of species diversity. *Ecology* **60**, 1225–1239.

Stebbins, G. L. (1965). Colonizing species of the native California flora. In *The Genetics of Colonizing Species* (ed. H. G. Baker and G. L. Stebbins), pp. 173–195. Academic Press, New York.

Sugihara, G., Grenfell, B. and May, R. M. (1990). Distinguishing error from chaos in ecological time series. *Philosophical Transactions of the Royal Society of London, Series B* **330**, 235–251.

Tainaka, K. (1988). Lattice model for the Lotka–Volterra system. *Journal of the Physical Society of Japan* **57**, 2588–2590.

Teramoto, E. (1993). Random disturbance and diversity of competitive systems. *Journal of Mathematical Biology* **31**, 761–769.

Tilman, D., May, R. M., Lehman, C. L. and Nowak, M. (1994). Habitat destruction and the extinction debt. *Nature* **371**, 65–66.

Tsuzuki, H. and Isogawa, Y. (1976). The first record of the rice water weevil, tentatively identified as *Lissorhoptrus oryzophilus*, in Aitl prefecture, Japan. *Syokubutu Boeki* **30**, 341.

Utida, S. (1957). Population fluctuation, an experimental and theoretical approach. *Cold Spring Harbour Laboratory, Symposia on Quantitative Biology* **22**, 139–151.

Udvardy, M. D. F. (1969). *Dynamic Zoogeography*. Van Nostrand Reinhold, New York.

Ulbrich, J. (1930). *Die Bisamratte*. Heinrich, Dresden.

Usher, M. B. (1989). Ecological effects of controlling invasive terrestrial vertebrates. In *Biological Invasions: A Global Perspective* (ed. J. A. Drake, H. A.

Mooney, F. di Castri, R. H. Groves, F. J. Kruger, M. Rejmánek and M. Williamson), SCOPE 37, pp. 463–489. Wiley, New York.

Van den Bosch, F., Metz, J. A. J. and Diekmann, O. (1990). The velocity of spatial population expansion. *Journal of Mathematical Biology* **28**, 529–565.

Van den Bosch, F., Hengeveld, R. and Metz, J. A. J. (1992). Analysing the velocity of animal range expansion. *Journal of Biogeography* **19**, 135–150.

Volpert, A. I., Volpert, V. A. and Volpert, V. A. (1994). *Traveling Wave Solutions of Parabolic Systems*. AMS, Providence, RI.

Wandeler, A., Capt, S., Kappeler, A. and Hauser, R. (1988). Oral immunization of wildlife against rabies: concept and first field experiments. *Reviews of Infectious Diseases* **10**, (Suppl. 4), S649–S653.

Wilcove, D. S., McLellan, C. H. and Dobson, A. P. (1986). Habitat fragmentation in the temperate zone. In *Conservation Biology: The Science of Scarcity and Diversity* (ed. M. E. Soule), pp. 237–256. Sinauer, Sunderland, MA.

Williamson, M. (1989). Mathematical models of invasion. In *Biological Invasions: A Global Perspective* (ed. J. A. Drake, H. A. Mooney, F. di Castri, R. H. Groves, F. J. Kruger, M. Rejmánek and M. Williamson), SCOPE 37, pp. 329–350. Wiley, New York.

Williamson, M. H. and Brown, K. C. (1986). The analysis and modelling of British invasion. *Philosophical Transactions of the Royal Society of London, Series B* **314**, 505–522.

Wing, L. (1943). Spread of the starling and English sparrow. *Auk* **60**, 74–78.

Wissel, C., Stephan, T. and Zaschke, S.-H. (1994). Modelling extinction and survival of small populations. In *Minimum Animal Populations* (ed. H. Remmert), Ecological Studies, vol. 106, pp. 67–104. Springer-Verlag, Berlin.

Yachi, S., Kawasaki, K., Shigesada, N. and Teramoto, E. (1989). Spatial patterns of propagating waves of fox rabies. *Forma* **4**, 3–12.

Index